World Conference on Carbon (CARBON 2018)

Bonded by Carbon

Madrid, Spain
1-6 July 2018

Editors:

Alonso Morales
Carrasco Marín
Lázaro Elorri
María Jesus

ISBN: 978-1-51087885-3

Printed from e-media with permission by:

Curran Associates, Inc.
57 Morehouse Lane
Red Hook, NY 12571

Some format issues inherent in the e-media version may also appear in this print version.

Printed by Curran Associates, Inc. (2019)

Spanish Carbon Group (GEC)
Instituto de Carboquimica, ICB-CSIC
C/ Miguel Luesma Castan
4 · 50018 · Zaragoza · Spain

http://www.gecarbon.org/

Additional copies of this publication are available from:

Curran Associates, Inc.
57 Morehouse Lane
Red Hook, NY 12571 USA
Phone: 845-758-0400
Fax: 845-758-2633
Email: curran@proceedings.com
Web: www.proceedings.com

Keynotes Index

Page:

Keynotes Index

Keynotes Index

Bonded by Carbon

Carbon2018

The World Conference on Carbon

Madrid, SPAIN, July 1st-6th

Book of Abstracts

www.carbon2018.org

Carbon 2018

The World Conference on Carbon

July 1st (Sun) - 6th (Fri), 2018

Madrid Marriott Auditorium Hotel & Conference Center
Madrid, Spain

Editores:
Alonso Morales, Noelia; Universidad Autónoma de Madrid

Carrasco Marín, Francisco; Universidad de Granada

Lázaro Elorri, María Jesus; Instituto de Carboquímica, CSIC

Sponsors

WELCOME TO CARBON 2018, WELCOME TO SPAIN

The Spanish Carbon Group (GEC) has the pleasure to invite you to the World Conference on Carbon.

On behalf of the Spanish Carbon Group (GEC) and the European Carbon Association, I would like to cordially welcome you in Madrid for the World Conference on Carbon 2018 "Bonded by Carbon". The Spanish Carbon Group had the privilege of organising Carbon 1994 in Granada, Andalusia, and Carbon 2003 in Oviedo, Asturias. This year's Carbon Conference is in Madrid, the capital of Spain. Madrid is a vibrant and cosmopolitan city where you will feel at home. Its cultural and leisure opportunities will guarantee you a great stay during the conference.

The tradition of carbon meetings dates back to 1953, in Buffalo, USA, under the leadership of Professor Stach Mrozowski, when four papers were presented. The World Conference on Carbon has therefore come a long way and, in the present edition, more than 800 abstracts have been received, 49 of which will be presented as keynotes, 360 as oral presentations and more than 400 as posters. Additionally, five awards will be granted. In Carbon 2018, we have made a great effort to increase the number of oral presentations. In fact, we have arranged up to six parallel rooms for two days to give the floor to as many speakers as possible. Plenary lectures will be delivered by Prof. Diego Cazorla, Prof. Yuri Gogotsi, Prof. Rosa Menéndez, Prof. Maurizio Prato and Prof. Hui-Ming Cheng. Moreover, on the occasion of the conference celebration, three different satellite courses will be held. Besides, some of the communications presented may be published in one of the special issues of the journals Carbon, Electrochimica Acta and Nanomaterials.

We will receive more than 700 participants from 55 different countries of all continents, more than 200 being students. The conference provides a great opportunity for interaction between more experienced researchers and the younger generation entering the world of carbon.

The Conference would not have been possible without the effort of all the colleagues involved in the Organizing and the Scientific Committees, to whom speccial thanks are in order. Furthermore, we are grateful to our sponsors for the financial support.

We wish all the participants an unforgettable week in Madrid and we hope you have time to discuss and interact with one another, as well as to visit the city.

María Jesús Lázaro
President of the Spanish Carbon Group
Chair of the European Carbon Association

Bonded by Carbon
Carbon2018
The World Conference on Carbon

COMMITTEES

Organizing Committee (Board of Spanish Carbon Group)

Chair

- María Jesus Lázaro Elorri (Instituto de Carboquímica, CSIC, Zaragoza)

- Francisco Carrasco Marín (Universidad de Granada)
- José Rodríguez Mirasol (Universidad de Málaga)
- Dolores Lozano Castelló (Universidad de Alicante)
- Francisco José Maldonado Hódar (Universidad de Granada)
- Fabián Suarez García (Instituto Nacional del Carbón, CSIC, Oviedo)
- Miguel Ángel Álvarez Merino (Universidad de Jaén)
- Isabel Suelves Laiglesia (Instituto de Carboquímica, CSIC, Zaragoza)
- María Olga Guerrero Pérez (Universidad de Málaga)
- Miguel Angel Gilarranz Redondo (Universidad Autónoma de Madrid)
- Mª Ángeles Lillo Ródenas (Universidad de Alicante)
- Agustín Francisco Pérez Cadenas (Universidad de Granada)
- Covadonga Pevida García (Instituto Nacional del Carbón, CSIC, Oviedo)
- Manuel Sánchez Polo (Universidad de Granada)

Local Organizing Committee

Chairs

- Juan José Rodríguez Jiménez (Universidad Autónoma de Madrid, Spain)
- Miguel Angel Gilarranz Redondo (Universidad Autónoma de Madrid, Spain)

Treasurer

- Francisco Heras Muñoz (Universidad Autónoma de Madrid, Spain)

- José Palomar Herrero (Universidad Autónoma de Madrid, Spain)
- Luisa Calvo Hernandez (Universidad Autónoma de Madrid, Spain)
- Monserrat Tobajas Vizcaíno (Universidad Autónoma de Madrid, Spain)
- Ángel Fernández Mohedano (Universidad Autónoma de Madrid, Spain)
- Fabián Suarez García (Instituto Nacional del Carbón, CSIC, Oviedo, Spain)
- Luisa Gómez Sainero (Universidad Autónoma de Madrid, Spain)
- Isabel Suelves Laiglesia (Instituto de Carboquímica, CSIC, Zaragoza, Spain)
- María Olga Guerrero Pérez (Universidad de Málaga, Spain)
- Zahara Martínez de Pedro (Universidad Autónoma de Madrid, Spain)
- Ariadna Álvarez Montero (Universidad Autónoma de Madrid, Spain)
- Elena Díaz Nieto (Universidad Autónoma de Madrid, Spain)
- Asunción Quintanilla Gómez (Universidad Autónoma de Madrid, Spain)
- Jesús Lemus Torres (Universidad Autónoma de Madrid, Spain)
- Macarena Muñoz García (Universidad Autónoma de Madrid, Spain)
- Ana Bahamonde Santos (Universidad Autónoma de Madrid, Spain)

Program Committee

Chair

- Francisco Carrasco Marín (Universidad de Granada, Spain)

Program Committee Coordinator

- Noelia Alonso Morales (Universidad Autónoma de Madrid, Spain)

- Jorge Bedia García Matamoros (Universidad Autónoma de Madrid, Spain)
- Carolina Belver Coldeira (Universidad Autónoma de Madrid, Spain)
- Agustín Francisco Pérez Cadenas (Universidad de Granada, Spain)
- José Rodríguez Mirasol (Universidad de Málaga, Spain)
- Dolores Lozano Castelló (Universidad de Alicante, Spain)
- Francisco José Maldonado Hódar (Universidad de Granada, Spain)
- Fabián Suarez García (Instituto Nacional del Carbón, CSIC, Oviedo, Spain)
- Miguel Ángel Álvarez Merino (Universidad de Jaén, Spain)
- Isabel Suelves Laiglesia (Instituto de Carboquímica, CSIC, Zaragoza, Spain)
- María Olga Guerrero Pérez (Universidad de Málaga, Spain)
- Fernando Rubiera González (Instituto Nacional del Carbón, CSIC, Oviedo, Spain)
- Juan Ignacio Paredes Nachón (Instituto Nacional del Carbón, CSIC, Oviedo, Spain)
- Marcos Granda Ferreira (Instituto Nacional del Carbón, CSIC, Oviedo, Spain)
- Francisco Javier Narciso Romero (Universidad de Alicante, Spain)
- Ignacio Martin Gullón (Universidad de Alicante, Spain)
- Isabel Suarez Ruiz (Instituto Nacional del Carbón, CSIC, Oviedo, Spain)
- María Antonia Diez Díaz-Estébanez (Instituto Nacional del Carbón, CSIC, Oviedo, Spain)
- Francisco García Labiano (Instituto de Carboquímica, CSIC, Zaragoza, Spain)
- Ana Arenillas de la Puente (Instituto Nacional del Carbón, CSIC, Oviedo, Spain)
- María Victoria López Ramón (Universidad de Jaén, Spain)
- Emilia Morallon Nuñez (Universidad de Alicante, Spain)
- Ana Beatriz García Suarez (Instituto Nacional del Carbón, CSIC, Oviedo, Spain)
- Tomas Cordero Alcántara (Universidad de Málaga, Spain)
- Maria Teresa Izquierdo Pantoja (Instituto de Carboquímica, CSIC, Zaragoza, Spain)
- José Angel Menéndez Diaz (Instituto Nacional del Carbón, CSIC, Oviedo, Spain)
- Clara Blanco Rodriguez (Instituto Nacional del Carbón, CSIC, Oviedo, Spain)
- Mar Garcia Hernández (Instituto de Ciencia de Materiales de Madrid, CSIC, Spain)
- María Victoria Martínez Huerta (Instituto de Catálisis y Petroleoquímica, CSIC, Madrid, Spain)
- M. Ángeles Lillo (Universidad de Alicante, Spain)
- Manuel Sanchez Polo (Universidad de Granada, Spain)
- Covadonga Pevida García (Instituto Nacional del Carbón, CSIC, Spain)

International Committee

- Elżbieta Frąckowiak (Poznan University of Technology, Poland)
- Roland Pellenq (Centre Interdisciplinaire de Nanoscience de Marseille, France)
- Rodney Taylor (Orion Engineered Carbons, USA)
- Matthew Weisenberger (University of Kentucky, USA)
- Rodney Andrews (University of Kentucky, USA)
- Juan Matos Lale (Universidad de Concepción, Chile)

- Juan Carlos Moreno Pirajan (Universidad de los Andes, Colombia)
- Jose Rene Rangel Mendez (Instituto Potosino de Investigación Científica y Tecnológica, Mexico)
- Yuan Chen (The University of Sydney, Australia)
- Robert Hurt (Brown University, USA)
- Alberto Bianco (Centre National de la Recherche Scientifique, Strasbourg, France)
- Yongsheng Chen (Nankai University, China)
- Hiroaki Hatori (National Institute of Advanced Industrial Science and Technology, Japan)
- Gareth Neighbour (Birmingham City University, UK)
- Chong Rae Park (Seoul National University, Korea)
- Juan M. D. Tascon (Instituto Nacional del Carbón (CSIC), Spain)
- Mauricio Terrones (Pennsylvania State University, USA)
- Jesus Santamaría (Universidad de Zaragoza, Spain)
- Chun-Zhu Li (Curtin University of Technology, Australia)
- Ladislav Kavan (Charles University, Czech Republic)
- Elena Obraztsova (Russian Academy of Sciences, Russian Federation)
- Ljubisa Radovic (Pennsylvania State University, USA)
- Teresa Bandosz (The City College of New York, USA)
- François Béguin (Poznan University of Technology, Poland)
- Francisco Rodriguez-Reinoso (Universidad de Alicante, Spain)
- Carmen Clemente Jul (Universidad Politécnica de Madrid, Spain)
- José Rivera Utrilla (Universidad de Granada, Spain)
- Morinobu Endo (Shinshu University, Japan)
- Katsumi Kaneko (Shinshu University, Japan
- Takashi Kyotani (Sendai-Shi, Japan)
- José Juan Pis Martínez (Instituto Nacional del Carbón (CSIC), Spain)
- Irene Suarez-Martinez (Curtin University, Australia)
- Luiz Depine de Castro (Centro Tecnológico do Exército, Brasil)
- Malcolm Heggie (Loughborough University, UK)
- Michael Strano (Massachusetts Institute of Technology, USA)
- Carlos Moreno Castilla (Universidad de Granada, Spain)
- Rafael Moliner Álvarez (Instituto de Carboquímica (CSIC), Spain)
- Jose Luis Figueiredo (University of Porto, Portugal)
- Toshiaki Enoki (Tokyo Institute of Technology, Japan)
- Nestor Tancredi (Universidad de la República, Uruguay)
- Jose Zagal (Universidad de Santiago de Chile, Chile)
- Freek Kapteijn (Delft University of Technology, The Netherlands)
- Vicente Gómez Serrano (Universidad de Extremadura, Spain)
- Juan Adanez Elorza (Instituto de Carboquímica (CSIC), Spain)
- Roberto Arce (Instituto de Física del Litoral, Argentina)
- Corina Hebestreit (European federation of mining industries, Belgium)
- Luiz C. Pardini (Instituto de Aeronáutica e Espaço São José dos Campos, Brasil)
- Dionysios D Dionysiou (University of Cincinnati, USA)
- Ashutosh Sharma (Indian Institute of Technology Kanpur, India)

Program

Plenary speakers

DIEGO CAZORLA AMORÓS, UNIVERSIDAD DE ALICANTE, SPAIN

"Surface (electro)chemistry of carbon materials. About the role of oxygen and nitrogen heteroatoms"

Dr. Diego Cazorla-Amorós is Professor in Inorganic Chemistry at the University of Alicante and Director of the Materials Institute of this university. He graduated in Chemistry from the University of Alicante in 1988 and obtained his PhD from this University in 1991. He spent pre- and postdoctoral stays at the Université Claude Bernard (France), Tohoku University (Japan) and Brown University (USA), among others. He became Associate Professor in 1994 at the University of Alicante and promoted as Full Professor in 2002. He is co-author of more than 230 papers and book chapters and 13 patents and has supervised 22 PhD students. His research interests include the design of a wide range of materials, including carbon materials, zeolites, mesoporous materials and metal-containing nanoparticles. He is especially interested in the synthesis, characterization and applications of different forms of carbon materials. He has made the most relevant contributions in the synthesis of high surface area porous carbons and in the modification of the surface chemistry in order to optimize their applications in the areas of environmental protection and energy storage and production.

YURY GOGOTSI, DREXEL UNIVERSITY, USA

"Transformative synthesis of carbon-based materials - from graphene to nanotubes and MXenes"

Dr. Yury Gogotsi is Charles T. and Ruth M. Bach Professor, Distinguished University Professor and Trustee Chair of Materials Science and Engineering at Drexel University. He also serves as Director of the A.J. Drexel Nanomaterials Institute. He received his MS (1984) and PhD (1986) from Kiev Polytechnic and a DSc degree from the Ukrainian Academy of Sciences in 1995. His research group works on nanostructured carbons, 2D carbides and other nanomaterials for energy, water and biomedical applciations. He has co-authored 2 books, 16 book chapters, more than 500 papers in peer-reviewed journals, edited 13 books, and obtained more than 50 patents. He was recognized as Highly Cited Researcher by Thomson-Reuters in 2014-2016 (h-index >100, Google Scholar).

He has received numerous awards for his research including the European Carbon Association Award, S. Somiya Award from the International Union of Materials Research Societies, International Nanotechnology Prize (RUSNANOPrize), R&D 100 Award from R&D Magazine (twice) and two Nano 50 Awards from NASA Nanotech Briefs. He has been elected a Fellow of the American Association for Advancement of Science (AAAS), Materials Research Society, American Ceramic Society, the Electrochemical Society, Royal Society of Chemistry, NanoSMAT Society, as well as Academician of the World Academy of Ceramics and Full Member of the International Institute for the Science of Sintering. He also serves on the MRS Board of Directors.

ROSA MENÉNDEZ, INSTITUTO NACIONAL DEL CARBÓN (CSIC), SPAIN

"State of the art and future perspectives of traditional carbon materials and new carbon forms"

Graduate and PhD in Chemistry. Full research professor at the Department of Chemistry of Materials, Instituto Nacional del Carbón (INCAR), CSIC in Oviedo, Spain. She has supervised 20 PhD theses and 22 MSc in the fields of Materials, Chemistry and Energy. She has published more than 200 papers in recognized journals and 9 patents. Main researcher in 28 projects, 10 of which belonged to European Research Programs, coordinator of 5. Awarded with recognitions such as: Schunk Carbon Award, XIX Science DuPont Award and the Spanish Materials Society to the best scientific career, among others. On parallel, she has occupied several leading positions such as director of INCAR and Vicepresident of CSIC. Active in National and International Committees, coordinator and member of several evaluation panels.

MAURIZIO PRATO, UNIVERSITÀ DI TRIESTE, ITALY

"Synthesis, Properties and Applications of Functionalized Carbon Nanoforms"

Maurizio Prato graduated in Padova, where he was appointed Assistant Professor in 1983. He moved to Trieste in 1992 as an Associate Professor to become Full Professor in 2000. From 2015, he is also Ikerbasque Research Professor at CIC BiomaGUNE, Spain. He spent sabbatical terms at Yale University with Sam Danishefsky (1986-7) and at the University of California, Santa Barbara with Fred Wudl (1991-2). He was Visiting Professor at the École Normale Supérieure de Paris (2001), at the University of Namur, Belgium (2010) and the University of Strasbourg (2014). He was the recipient of an ERC Advanced Research Grant, European Research Council, in 2008 and became a Member of the National Academy of Sciences (Accademia Nazionale dei Lincei) in 2010.

His research focuses on the synthesis and applications of new functional materials, for the study of their optical and electrochemical properties. Especially, he has worked on the functionalization of carbon nanostructures, including fullerenes, carbon nanotubes and graphene. He developed a series of organic reactions that make these materials more biocompatible, less or even non toxic, amenable to further functionalization, and easier to manipulate. Presently, he is interested in transferring the properties of these materials into applications, which include the possibility of spinal cord repair, the splitting of water, and the reduction of carbon dioxide into useful chemicals.

HUI-MING CHENG SHENYANG
NATIONAL LABORATORY FOR MATERIALS SCIENCE (CAS), CHINA

"Graphene Materials: Fabrication and Use in Electrochemical Energy Storage"

Dr. Hui-Ming Cheng is Professor and Director of both Advanced Carbon Research Division of Shenyang National Laboratory for Materials Science, Institute of Metal Research, the Chinese Academy of Sciences, and the Low-Dimensional Material and Device Laboratory of the Tsinghua-Berkeley Shenzhen Institute, Tsinghua University. His research activities focus on carbon nanotubes, graphene, other low-dimensional materials, energy storage materials, photocatalytic semiconducting materials, and bulk carbon materials. He has published over 500 papers with >50,000 citations and is recognized as a Highly Cited Researcher in both materials science and chemistry fields by Thomson Reuters. He used to be the Editor of Carbon and Editor-in-Chief of New Carbon Materials, and is now the founding Editor-in-Chief of Energy Storage Materials and Associate Editor of Science China Materials. He was elected as a member of Chinese Academy of Sciences and a fellow of TWAS.

Program at a glance

	SUNDAY 1ST	MONDAY 2ND					TUESDAY 3RD					
09:00 - 9:30		**OPEN CEREMONY**					**Plenary Lecture** **YURY GOGOTSI** Drexel University, USA					
09:30 - 10:00		**Plenary Lecture** **ROSA MENENDEZ** Instituto Nacional del Carbón (CSIC) Spain										
10:00 - 10:30							**COFFEE BREAK**					
10:30 - 10:45		**COFFEE BREAK**					**T03** (R5)	**T02** (R4)	**T05** (R2)	**T06** (R1)	**T07** (R6)	**T08** (R3)
10:45 - 11:00												
11:00 - 11:15		**T02** (R4)	**T04** (R5)	**T06** (R1)	**T07** (R3)	**T09** (R2)						
11:15 - 11:30												
11:30 - 11:45												
11:50 - 12:20		**T02-K** (R4)	**T04-K** (R5)	**T06-K** (R1)	**T07-K** (R3)	**T09-K** (R2)	**T01-K** (R5)	**T02-K** (R4)	**T05-K** (R2)	**T06-K** (R1)	**T07-K** (R6)	**T08-K** (R3)
12:20 - 12:35		**T02** (R4)	**T04** (R5)	**T06** (R1)	**T07** (R3)	**T09** (R2)	**T03** (R5)	**T02** (R4)	**T05** (R2)	**T06** (R1)	**T07** (R6)	**T08** (R3)
12:35 - 12:50												
12:50 - 13:05												
13:05 - 13:20												
13:20 - 13:35												
13:35 - 15:00		**LUNCH**					**LUNCH**					
15:15		**T02** (R4)	**T04** (R5)	**T06** (R1)	**T07** (R3)	**T09** (R2)	**T01** (R5)	**T02** (R4)	**T05** (R2)	**T06** (R1)	**T07** (R6)	**T08** (R3)
15:30												
15:45			**T03** (R5)									
16:00	**REGISTRATION** (16:00 to 20:00) (Madrid Marriott Auditorium & Conference Center)											
16:15												
16:30 - 17:00		**T02-K** (R4)	**T04-K** (R5)	**T06-K** (R1)	**T07-K** (R3)	**T09-K** (R2)	**T01-K** (R5)	**T02-K** (R4)	**T05-K** (R2)	**T06-K** (R1)	**T07-K** (R6)	**T08-K** (R3)
17:00 - 17:30		**Poster Session & Coffee** (T02, T04, T06, T09)					**Poster Session & Coffee** (T01, T03, T06, T07)					
17:30 - 19:00												
19:00 - 19:30							**SOCIAL EVENT** (Madrid Guided Tour)					
19:30 - 20:00	**WELCOME COCKTAIL**											
20:00												

WEDNESDAY 4TH	THURSDAY 5TH	FRIDAY 6TH
Plenary Lecture **HUI-MING CHEN** Shenyang National Lab for Materials Science (CAS), China	**Plenary Lecture** **DIEGO CAZORLA-AMORÓS** Universidad de Alicante, Spain	**Plenary Lecture** **MAURIZIO PRATO** Università di Trieste, Italy
COFFEE BREAK	**COFFEE BREAK**	**COFFEE BREAK**
T01 (R5) · **T02** (R4) · **T05** (R2) · **T06** (R6) · **T08** (R3)	**T01** (R5) · **T02** (R4) · **T05** (R2) · **T06** (R1) · **T08** (R3)	**T01** (R5) · **T05** (R2) · **T08** (R3) · **T10** (R4)
T01-K (R5) · **T02-K** (R4) · **T05-K** (R2) · **T06-K** (R1 & R6) · **T08-K** (R3)	**T01-K** (R5) · **T02-K** (R4) · **T05-K** (R2) · **T06-K** (R1) · **T08-K** (R3)	**T01-K** (R5) · **T05-K** (R2) · **T06-K** (R1) · **T08-K** (R3) · **T10-K** (R4)
T01 (R5) · **T02** (R4) · **T05** (R2) · **T06** (R1 & R6) · **T08** (R3)	**T01** (R5) · **T02** (R4) · **T05** (R2) · **T06** (R1) · **T10** (R3)	**T01** (R5) · **T05** (R2) · **T06** (R1) · **T08** (R3) · **T10** (R4) **AWARD SESSION** (R1) Sustainable Energy & Fuels Energy & Environmental Science
LUNCH	**LUNCH**	**CLOSING CEREMONY** 14:00 **LUNCH**
T01 (R5) · **T02** (R4) · **T05** (R2) · **T06** (R1 & R6) · **T08** (R3)	**T01** (R5) · **T02** (R4) · **T05** (R2) · **T06** (R1) · **T08** (R3)	
T01-K (R5) · **T02-K** (R4) · **T05-K** (R2) · **T06-K** (R1 & R6) · **T08-K** (R3)	**T01-K** (R5) · **T02-K** (R4) · **T05-K** (R2) · **T06-K** (R1) · **T10-K** (R3)	
Poster Session & Coffee (T05, T08, T10)	**COFFEE BREAK** **AWARD SESSION**	
	CONFERENCE DINNER (Casa de Mónico)	

- T01_Graphene and graphite.
- T02_Advanced carbon materials: nanotubes, fullerenes, carbon fibres and carbon based composites.
- T03_Coal, coke and carbon black.
- T04_ Thermo-chemical conversion processes.
- T05_Preparation, characterization, simulation and modelling of carbon.
- T06_Electrochemical applications.
- T07_Catalysis application.
- T08_Environmental applications.
- T09_Gas separation and storage.
- T10_Carbons for health, medical and biological applications.

Oral Index

Oral Index

Code	Title
0006	THE INFLUENCE OF THE FLOW PATTERN OF THE CONTAMINATED AIR ON THE ADSORPTION BEHAVIOUR OF AN ACTIVATED CARBON FILTER. LODEWYCKX, PETER - WINDEY, BRECHT - VAN RIET, ROMUALD
0015	CU2O ON 1-OCTYL-3-METHYLIMIDAZOLIUM CHLORIDE ASSISTED GRAPHITE SHEETS AS EFFICIENT ELECTROCATALYST FOR CO2 REDUCTION NING, HUI - WANG, WENHANG - MAO, QINHU - WU, MINGBO
0023	ELECTROTHERMAL VAPORIZATION–INDUCTIVELY COUPLED PLASMA-MASS SPECTROMETRY OF ISOSTATICALLY PRESSED FINE GRAIN CARBON JOKANOVIC, BOJAN - LIERFELD, THOMAS - LATZEL, SABINE - CHODURA, ALFRED
0025	BOTTOM-UP DESIGN OF NITROGEN-CONTAINING CARBON CATALYSTS FOR OXYGEN REDUCTION REACTION SHIBUYA, RIKU - KONDO, TAKAHIRO - NAKAMURA, JUNJI
0029	DIFFERENT IMPACT OF REDUCED GRAPHENE OXIDE ON PHOTOCATALYTIC WATER SPLITTING SYSTEMS WOHLRAB, SEBASTIAN - SONNECK, MARCEL
0030	SINGLE WALLED CARBON NANOTUBE (SWCNT) BASED STRETCHABLE STRAIN SENSOR AHUJA, PREETY - KUKOBAT, RADOVAN - KANEKO, KATSUMI
0031	STRUCTURE-PERFORMANCE RELATIONSHIPS OF CARBON-SUPPORTED IRON NANOPARTICLES IN THE FISCHER-TROPSCH SYNTHESIS OF SHORT OLEFINS OSCHATZ, MARTIN - LAMA, SANDY - ANTONIETTI, MARKUS - DE JONG, KRIJN
0032	ELECTROCHEMICALLY EXFOLIATED GRAPHENE: PRODUCTION, HYBRIDIZATION AND ELECTROCHEMICAL APPLICATION WANG, JUNZHONG - WANG, JUNYING
0033	NEW PROCESS FOR EFFICIENT CO2 CAPTURE BY INNOVATIVE ADSORBENTS BASED ON MODIFIED GRAPHENE AEROGELS AND MOF MATERIALS BENEDITO BORRÁS, ADOLFO
0034	TOWARDS BETTER UNDERSTANDING OF ION STORAGE/DIFFUSION MECHANISM IN SUPERCAPACITORS BASED ON HIERARCHICAL POROUS CARBONS YAN, RUNYU - OSCHATZ, MARTIN - ANTONIETTI, MARKUS
0035	INTERFACIAL CHARGE TRANSFER IN GRAPHENE-BASED 'HYBRIDS' WITH TRANSITION METAL OXIDES USING IN-SITU RAMAN SPECTROSCOPY GUPTA, SANJU
0040	CARBON NANOSPHERES FOR IN-SITU IMPROVEMENT OF HEAVY CRUDE OILS IN OXIDIZING AND INERT ATMOSPHERES BETANCUR MARQUEZ, STEFANIA - FRANCO ARIZA, CAMILO ANDRES - MALDONADO HODAR, FRANCISCO JOSÉ - CARRASCO MARIN, FRANCISCO - CORTES CORREA, FARID BERNARDO
0044	CARBON AEROGELS FOR ELECTROCHEMICAL APPLICATIONS CANAL RODRÍGUEZ, MARÍA - REY RAAP, NATALIA - ARENILLAS PUENTE, ANA - MENÉNDEZ DÍAZ, JOSE ÁNGEL
0047	ROOM TEMPERATURE SYNTHESIS OF GRAPHITIC POROUS CARBON NANOSTRUCTURES CASCO, MIRIAN ELIZABETH - TOLOSA, AURA - PRESSER, VOLKER - BADACZEWSKI, FELIX - SMARSLY, BERND - BORCHARDT, LARS
0050	TRUNCATED OCTAHEDRAL TIO2/FE2O3 NANOPARTICLES WITH GRAPHENE SUPPORTS FOR WATER TREATMENT APPLICATIONS YURÜM, ALP - GOHARI BAJESTANI, ZAHRA - YURUM, YUDA
0052	FUNCTIONALIZATION OF FEW LAYER GRAPHENE FOR DESIGNING HETEROGENEOUS CATALYSTS BASED ON ORGANOMETALLIC COMPOUNDS NÁJERA-DOMINGO, CARMEN - SANSANO-GIL, JOSÉ MIGUEL - FERRÁNDIZ-SAPERAS, MARCOS - GHISOLFI, ALESSIO - CAZORLA-AMORÓS, DIEGO
0053	NANOCASTING IN BALL MILLS - COMBINING ULTRA-HYDROPHILICITY AND ORDERED MESOPOROSITY IN CARBON MATERIALS EN, ZHANG - LARS, BORCHARDT - STEFAN, KASKEL
0055	CCVD SYNTHESIS OF GRAPHENE RELATED MATERIALS USING STAINLESS STEEL FOAMS AS STRUCTURED CATALYSTS MONZON, ANTONIO - TARIFA, PILAR - SEBASTIÁN, VÍCTOR - LATORRE, NIEVES - ROMEO, EVA - CAZAÑA, FERNANDO
0056	MECHANICAL BEHAVIOR OF 3D CARBON/CARBON COMPOSITES : THE ROLE OF INTERNAL INTERFACES GILLARD, ADRIEN P. - COUEGNAT, GUILLAUME - CHUPIN, SYLVAIN - VIGNOLES, GERARD L.
0057	CONVERSION OF ORGANIC CRYSTALS INTO ORDERED CARBONACEOUS FRAMEWORKS WITH MICROPOROSITY NISHIHARA, HIROTOMO - MATSUURA, KENTA - OHWADA, MAO - MATSUO, YOSHIAKI - MARUYAMA, JUN - HAYASAKA, YUICHIRO - YAMAGUCHI, SHINGI - KAMIYA, KAZUHIDE - TANI, FUMITO - KYOTANI, TAKASHI
0058	PYROLYSIS STUDY OF OIL PALM SHELLS WITH THE NOT REACTED CORE MODEL GÁMEZ, SEBASTIÁN - DE LA TORRE, ERNESTO
0059	BIO-BASED ELECTRODE MATERIALS FOR ENERGY STORAGE TECHNOLOGIES (ELECTROCHEMICAL DOUBLE-LAYER CAPACITORS) HOFFMANN, VIOLA - CORREA RODRIGUEZ, CATALINA - KRUSE, ANDREA

Code	Title
0060	MALACHITE GREEN DYE REMOVAL FROM WATER USING DATE PIT DERIVED ACTIVATED CARBON HIJAB, MOUHAMMAD - SALI, SAFAE - MACKEY, HAMISH - MCKAY, GORDON
0068	ANIONIC FUNCTIONALIZATION OF NANOCARBONS: A POWERFUL AND VERSATILE TOOL SHEET, DEBOBRATA - RIANT, OLIVIER - HERMANS, SOPHIE - ANTONIN, DESMECHT
0072	ABLATION BEHAVIOR OF C/C-ZRC-SIC COMPOSITES UNDER OXYACETYENE TORCH AT TWO HEAT FLUXES LI, KEZHI
0073	IN-SITU MEASUREMENT OF THE IN-PLANE STRUCTURE CHANGE OF GRAPHITE ANODE OF LI ION BATTERY DURING THE CHARGE/DISCHARGE PROCESS FUJIMOTO, HIROYUKI - TAKAGI, SHIGEHARU - KIUCHI, HISAO - NAKA, TAKAHIRO - MATSUBARA, EIICHIRO - UKYO, YOSHIO - ABE, TAKESHI - OGUMI, ZEMPACHI
0076	TURBOSTRATIC POLYHEDRAL CARBON NANO-ONIONS SELECTIVELY SYNTHESIZED BY ARC DISCHARGE IN WATER ALESSANDRO, FRANCESCA - SCARCELLO, ANDREA - BASANTES VALVERDE, MARLON DANILO - COELLO FIALLOS, DIANA CAROLI-NA - OSMAN, SALIH MOHAMMED - CUPOLILLO, ANNA - ARIAS, MELVIN - DE LUCA, GIORGIO - CURCIO, EFREM - CAPUTI, LORENZO SALVATORE
0079	CHARACTERIZATION OF CARBON PROPERTIES INFLUENCING ON BATTERY PERFORMANCE TAKAGI, SHIGEHARU - FUJIMOTO, HIROYUKI - KIUCHI, HISAO - NAKA, TAKAHIRO - MATSUNAGA, TOSHIYUKI - UKYO, YOSHIO - FUKUNAGA, TOSHIHARU - ABE, TAKESHI - OGUMI, ZEMPACHI - MATSUBARA, EIICHIRO
0080	EFFECTIVE ADSORPTIVE REMOVAL OF GALLIUM ON SYNTHESIZED MESOPOROUS CARBONS BERNABÉ VÍRSEDA, IGNACIO - SÁEZ GONZALEZ, PATRICIA - GÓMEZ MARTÍN, JOSÉ MARÍA - DÍEZ ALCÁNTARA, EDUARDO - RODRÍ-GUEZ RODRÍGUEZ, ARACELI
0082	DETERMINING MICROPORE SIZE DISTRIBUTIONS AND MECHANICAL PROPERTIES OF MICROPOROUS CARBONS USING ADSORPTION-INDUCED STRAINS BALZER, CHRISTIAN - REICHENAUER, GUDRUN
0083	TEMPERATURE DEPENDENCES OF COEFFICIENT OF THERMAL EXPANSION WITH DIFFERENT TYPE OF ARTIFICIAL GRAPHITE IWASHITA, NORIO
0084	ELECTRICAL DOUBLE LAYER CAPACITORS – FINDINGS FROM FUNDAMENTAL RESEARCH AND THEIR IMPACT ON THE DESIGN OF STORAGE DEVICES LORRMANN, VOLKER - WIENER, MATTHIAS - REICHENAUER, GUDRUN
0086	TEMPLATE-FREE SYNTHESIS OF POROUS “NOBLE” CARBON-NITRIDE MATERIALS WITH HIGH SURFACE POLARITY FOR SELECTIVE GAS ADSORPTION WALCZAK, RALF - ANTONIETTI, MARKUS - OSCHATZ, MARTIN
0090	PREPARATION AND CHARACTERIZATION OF BIO-BASED ACTIVATED CARBON MATERIALS FOR USE AS SUPERCAPACI-TOR ELECTRODES BREITENBACH, STEFAN - UNTERWEGER, CHRISTOPH - FÜRST, CHRISTIAN - HASSEL, ACHIM WALTER
0093	3D PRINTING OF HIERARCHICALLY STRUCTURED HIGH SURFACE AREA ACTIVATED CARBONS STELDINGER, HENDRYK - GLÄSEL, JAN - ETZOLD, BASTIAN J.M.
0095	THE „IN-SITU ELECTROLYTE“ CONCEPT – A HOLISTIC SYNTHESIS STRATEGY TOWARDS CARBON-BASED SUPERCAPACITORS LEISTENSCHNEIDER, DESIRÉE - SCHNEIDERMANN, CHRISTINA - TROSCHKE, ERIK - GRÄTZ, SVEN - BORCHARDT, LARS
0103	SYNTHESIS OF NANOCARBONACEOUS MATERIALS BY DECOMPOSITION OF METHANE USING CATALYSTS OF MN-ME(NI,CO)/CB. MONZON, ANTONIO - CAZAÑA, FERNANDO - SEBASTIÁN, VICTOR - LATORRE, NIEVES - ROMEO, EVA - ROYO, CARLOS - TARIFA, PILAR
0104	CARBON NANOTUBE CONTAINING POLYACRYLONITRILE MATERIALS FOR THE EFFICIENT OXYGEN EVOLUTION REAC-TION – A CONDUCTIVITY STUDY BEINE, ANNA KATHARINA - BROICHER, CORNELIA - HU, QINGTAO - MAYERL, LISA - LU, AN-HUI - PALKOVITS, REGINA
0105	TUNING ELECTRONIC PROPERTIES OF CARBON NANOTUBES VIA OXYGEN PLASMA – EXPERIMENTAL AND THEORETI-CAL APPROACH DUCH, JOANNA - MAZUR, MICHAL - GOLDA-CEPA, MONIKA - PISKORZ, WITOLD - KOTARBA, ANDRZEJ
0107	NITROGEN-DOPED BIOMASS-DERIVED CARBON BY MECHANOCHEMICAL SYNTHESIS FOR ELECTROCHEMICAL ENER-GY STORAGE SCHNEIDERMANN, CHRISTINA - KENSY, CHRISTIAN - GIEBELER, LARS - BORCHARDT, LARS
0110	PREPARATION AND ELECTROMAGNETIC WAVE ABSORBING PROPERTIES OF 3D GRAPHENE/PINE NEEDLE-LIKE IRON NANO-ACICULAR WHISKER COMPOSIT ZHAO, TINGKAI
0112	THE ELECTROCHEMICAL DECONSOLIDATION MECHANISM OF GRAPHITE MATRIX IN HTR SPHERICAL FUEL ELEMENTS CHEN, XIAOTONG - ZHANG, CHI - LU, ZHENMING - LIU, BING

Oral Index

Code	Title
0116	STRUCTURAL CHARACTERISATION OF HEAT-TREATED CARBON: FROM MOLECULE-LIKE TO CRYSTALLINE CARBON SCHUEPFER, DOMINIQUE B. - BADACZEWSKI, FELIX - GUERRA, JUAN-MANUEL - HOFMANN, DETLEV M. - SMARSLY, BERND M. - HEILIGER, CHRISTIAN - KLAR, PETER J.
0120	BIOACTIVE COATINGS ON CARBON FIBER FOR BONE TISSUE APPLICATION ZHANG, LEILEI
0123	MULTI-FUNCTIONAL CARBON FIBRES FOR HIGH-DENSITY ENERGY STORAGE IN LITHIUM-SULFUR-BATTERIES TOM, BOENKE - PAUL, HÄRTEL - SUSANNE, DÖRFLER - STEFAN, KASKEL
0125	RUTHENIUM SUPPORTED ON VARIOUS CARBON MATERIALS FOR SUGAR HYDROGENATION SANTOS, JOSE LUIS - AHO, ATTE - MURZIN, DMITRY - CENTENO, MIGUEL ANGEL - ODRIOZOLA, JOSE ANTONIO
0128	N-SITU SAXS STUDY OF EXFOLIATION AND REDUCTION PROCESSES OF GO POWDER AND AQUEOUS SUSPENSION SYSTEMS DURING HEATING HERNANDEZ BELMARES, PERLA JANETTH - RODRIGUEZ FERNANDEZ, OLIVERIO - HERNANDEZ HERNANDEZ, ERNESTO - GONZALEZ MORONES, PABLO - MATA PADILLA, JOSE MANUEL - GALLARDO VEGA, CARLOS A.
0130	TOWARDS SUSTAINABLE ENERGY STORAGE: RECENT DEVELOPMENTS IN DUAL-CARBON BATTERIES TOBIAS, PLACKE - ANDREAS, HECKMANN - KOLJA, BELTROP - MARTIN, WINTER
0132	DIFFERENTIAL RAMAN BACKSCATTERING CROSS SECTIONS OF BLACK CARBON NANOPARTICLES LE, KIM CUONG - LEFUMEUX, CHRISTOPHE - PINO, THOMAS
0133	THE ADVENT OF SOOT NANOSTRUCTURE AND SOOT EVOLUTION BY RAMAN, FAR-UV TO MID-IR SPECTROSCOPIES AND HRTEM LE, KIM CUONG - GAVILAN, LISSETH - PRÉ, PASCALINE - DELDICQUE, DAMIEN - DARTOIS, EMMANUEL - ROUZAUD, JEAN-NOËL - PINO, THOMAS
0135	SYNTHESIS AND CHARACTERIZATION OF CARBON-BASED ION IMPRINTED MATERIALS FOR SELECTIVE RECOGNITION AND RECOVERY OF LITHIUM ION LIU, WEIFENG - AN, ZHUOLIN - YAN, GUGANG - LIANG, QI - QIN, LEI - LIU, XUGUANG - YANG, YONGZHEN - XU, BINGSHE
0137	IMPROVED BRIGHTNESS OF CARBON DOT-BASED ELECTROLUMINESCENT LEDS VIA BLENDING PVK IN EMITTING LAYER XU, JIACONG - ZHENG, JINGXIA - DING, YUANFEI - MIAO, YANQIN - YANG, YONGZHEN - LIU, XUGUANG - JIA, HUSHENG
0139	THE FILMS DERIVED FROM SOL-GEL BASED SWCNT INKS KUKOBAT, RADOVAN - KANEKO, KATSUMI
0141	POLYACRYLONITRILE FIBERS CONTAINING THIN-LAYER GRAPHENE PARTICLES AS PRECURSORS FOR CARBON FIBERS MIRKO, RICHTER - NGUYEN HOAI AN, TRAN - ROLF-DIETER, HUND - CHERIF, CHERIF - WOLZ, DANIEL SEBASTIAN - ROMANÍ VÁZQUEZ, ADRIÁN - SHAYGAN NIA, ALI - LOHE, MARTIN R. - BÖHM, ROBERTO - FENG, XINLIANG - JÄGER, HUBERT
0145	MODIFIED CARBON XEROGEL CATALYSTS FOR SELECTIVE HYDROLYSIS OF BIOMASS: THE IMPACT OF SURFACE CHEMISTRY AND ACTIVE SITES MORAWA EBLAGON, KATARZYNA - MALAIKA, ANNA - R PEREIRA, MANUEL FERNANDO - FIGUEIREDO, JOSE LUIS
0146	THERMOCHEMICAL CONVERSION DURING STABILISATION OF LIGNIN/PAN BLENDED PRECURSOR AND THEIR EFFECT ON POROSITY BENJAMIN, RICHTER
0149	MICROSTRUCTURE AND ABLATION MECHANISM OF SIC-ZRC-AL2O3 COATING FOR SIC COATED C/C COMPOSITES UNDER OXYACETYLENE TORCH TEST GUO, LINGJUN
0150	ACTIVATED CARBON-BASED P-FE NANOCOMPOSITES FOR THE ADSORPTION AND PHOTODEGRADATION OF METHYLENE BLUE UNDER SOLAR IRRADIATION MATOS, JUAN - MONTAÑA, RICMARY - GONCALVES, GUSTAVO R. - SCHETTINO JR., MIGUEL A. - PASSAMANI, EDSON C. - FREITAS, JAIR C.C
0151	ZIFS-DERIVED NIXCO3-XO4/CNTS NANOCOMPOSITES WITH ENHANCED ELECTROCHEMICAL PERFORMANCE FOR SUPERCAPACITOR XUE, BEI
0154	AN EXPERIMENTAL AND MOLECULAR SIMULATION STUDY OF THE STRENGTHENING OF CNT FIBERS THROUGH GRAPHENE OXIDE INTERLOCKING LU, WEIBANG - QU, SHUXUAN - GONG, WENBIN - ZHOU, GENHENG - JIANG, JIN
0157	DIRECT RESISTIVE HEATING OF CARBON FIBERS: USE FOR CARBONIZATION/GRAPHITIZATION OF CARBON FIBER TEXTILES OR CFRP RECYCLING KLOIBER, VERENA - PAMPEL, JONAS - LEHMANN, BEATA - DIETZEL, YVETTE - DÖRFLER, SUSANNE - KASKEL, STEFAN - BEYER, ECKHARD
0165	EFFECT OF OXYGEN AND HYDROGEN ON MICROSTRUCTURE OF PYROLYTIC CARBON DEPOSITED FROM THERMAL DECOMPOSITION OF METHANE AND ETHAN REN, BIYUN

Code	Title
0168	THEORETICAL MODEL FOR STRUCTURE DETERMINATION OF HIERARCHICALLY STRUCTURED CARBON NANOTUBE YARN PARK, JUNBEOM - LEE, JAEGEUN - LEE, SUNG-HYUN - LEE, KUN-HONG - KIM, SEUNG MIN
0171	STRUCTURE AND DYNAMICS OF NANOCRYSTALLINE GRAPHENE AT HIGH TEMPERATURES REVEALED BY IN SITU TRANSMISSION ELECTRON MICROSCOPY SHYAM KUMAR, C.N. - CHAKRAVADHANULA, VENKATA SAI KIRAN - CHAKRAVADHANULA, VENKATA SAI KIRAN - DEHM, SIMONE - WANG, DI - KRUPKE, RALPH - KÜBEL, CHRISTIAN
0172	REACTIVITY OF FREE AND TRAPPED CHARGE CARRIERS OF PHOTOEXCITED NANOPOROUS CARBONS GOMIS-BERENGUER, ALICIA - LIMA, JOAO CARLOS - ANIA, CONCHI
0173	IMPROVED H2S AND HG REMOVAL FROM NATURAL GAS USING PETCOKE DERIVED ADSORBENTS ANIA, CONCHI - GARCIA-GONZALEZ, RAQUEL - RODRIGUEZ-TROBAJO, JAIME - L. FLORES-LOPEZ, SAMANTHA - CONCHESO, ALEJANDRO - VALERA, MARIA EUGENIA - PEÑA, JOWAR - LOPEZ-ANTON, MARIA ANTONIA - MONTES-MORAN, MIGUEL A. - MOREN-VIRGEN, M.ROSARIO - PARRA, JOSE B. - MARTINEZ-TARAZONA, ROSA M. - D´ELIA CAMACHO, LUIS FELIPE
0175	GRAPHENE FOR BIOAPPLICATIONS: PREPARATION, CYTOTOXICITY AND INTEGRATION IN 3D SCAFFOLDS VÁZQUEZ, ESTER
0176	NOVEL METHODS TO TAILOR TRANSPORT POROSITY OF CURED PHENOLIC RESINS AND DERIVED CARBON MATERIALS KOZYNCHENKO, OLEKSANDR - DIAS-AUNON, JOSE ANTONIO
0177	EFFECTS OF OXIDATION DEGREE ON ACTIVATED CARBON FIBER FELT FOR SUPERCAPACITOR ELECTRODE SALDANHA MARCUZZO, JOSSANO - CASTILHO RODRIGUES, ALINE - DA SILVA PINHEIRO, BARBARA - CUÑA, ANDRES - TADAO MATSUSHIMA, JORGE - RIBEIRO BALDAN, MAURICIO - FONTANA BATISTA, ALINE
0182	CATALYTIC SYNTHESIS OF MULTIWALLED CARBON NANOTUBES ON FOSSILIZED BIOSILICA SUBSTRATE NAZHIPKYZY, MERUYERT - ZHALGASBAIKYZY, AIGERIM - ZHAPAROVA, AIZHAN - NURGAIN, ARAILYM - MANSUROV, ZULKHAIR - TEMIRGALIYEVA, TOLGANAY - LESBAYEV, BAKHYTZHAN - PRIKHODKO, NIKOLAY - TULEPOV, MARAT
0185	FACILE SYNTHESIS OF NOVEL BOWL-LIKE HOLLOW CARBON SPHERES BY COMBINATION OF HYDROTHERMAL CARBONIZATION AND SOFT TEMPLATING QIN, MINGLI - ZHANG, ZILI - JIA, BAORUI - WU, HAOYANG
0186	HIGH-PERFORMANCE SUPERCAPACITORS FROM PHOSPHORUS, NITROGEN AND OXYGEN CO-DOPED SPHERICAL POLYMER-BASED CARBONS HUANG, CONGCONG - PUZIY, ALEXANDER M. - PODDUBNAYA, OLGA I. - HULICOVA-JURCAKOVA, DENISA - SOBIESIAK, MAGDALENA - GAWDZIK, BARBARA
0188	TRANSFORMATION OF POLYIMIDE COPOLYMER TO NANOPOROUS CARBON ALONG HEAT TREATMENT WITH PHOSPHORIC ACID SOBIESIAK, MAGDALENA - GAWDZIK, BARBARA - PODDUBNAYA, OLGA I. - PUZIY, ALEXANDER - SPRYNSKYY, MYROSLAW
0192	ELECTROCHEMICAL FUNCTIONALIZATION OF CARBON MATERIALS WITH NITROGEN AND PHOSPHORUS CONTAINING COMPOUNDS QUINTERO-JAIME, ANDRÉS FELIPE - GHISOLFI, ALESSIO - CAZORLA-AMORÓS, DIEGO - MORALLÓN-NÚÑEZ, EMILIA
0193	ELECTROCHEMICAL IMMUNOSENSOR BASED ON CARBON NANOTUBES DECORATED WITH GOLD NANOPARTICLES FOR THE DETECTION OF PSA QUINTERO-JAIME, ANDRES FELIPE - BERENGUER-MURCIA, ANGEL - CAZORLA-AMOROS, DIEGO - MORALLON-NUÑEZ, EMILIA
0194	PREPARATION OF COAL LIQUEFACTION RESIDUES-BASED CARBON NANOFIBER NONWOVEN FABRIC LI, XIAO - TIAN, XIAODONG - HE, YITING - LIU, WENHONG - SONG, YAN - LIU, ZHANJUN
0195	RAPID CO2 ACTIVATION OF CARBON GELS USING MICROWAVE HEATING MUKAI, SHIN - FUKUMITSU, DAISUKE - IWAMURA, SHINICHIROH - YOSHIKAWA, MASAAKI
0198	THERMAL CONTACT RESISTANCE OF VARIOUS CARBON NANOMATERIAL-BASED EPOXY COMPOSITES DEVELOPED FOR THERMAL INTERFACE APPLICATIONS ALI RAZA, MOHSIN - WESTWOOD, AIDAN - STIRLING, CHRIS
0201	CHARGE/DISCHARGE PROPERTIES OF SULFUR MOLECULES CONFINED IN 1D CARBON NANOSPACES URITA, KOKI - FUJIMORI, TOSHIHIKO - NOTOHARA, HIROO - MORIGUCHI, ISAMU
0203	HARD CARBON MATERIALS FOR SODIUM ION BATTERIES (NIBS) BEDA, ADRIAN - TABERNA, PIERRE-LOUIS - SIMON, PATRICE - GHIMBEU, CAMÉLIA
0205	UNDERSTANDING OF ORR BY EXAMINING CARBON-OXYGEN GASIFICATION REACTION AND ACTIVE SITES ON METAL AND HETEROATOMS FREE CARBONS GABE, ATSUSHI - RUIZ-ROSAS, RAMIRO - MORALLÓN, EMILIA - CAZORLA-AMORÓS, DIEGO
0206	ALKALI METAL HALIDE AQUEOUS ELECTROLYTE BATTERIES USING SINGLE-WALLED CARBON NANOTUBES KAWASAKI, SHINJI - LI, CANGHAO - KATO, NAO - ISHII, YOSUKE

Oral Index

Code	Title
0209	PORE STRUCTURE AND SURFACE CHEMICAL STATE OF MESOPOROUS CARBONS DERIVED FROM MGO TEMPLATE PROCESS ANZAI, MIZUHO
0210	PORE-SIZE SENSITIVE CO-ION ASSOCIATION OF IONIC LIQUIDS IN CARBON NANOPORES FUTAMURA, RYUSUKE - IIYAMA, TAKU - GOGOTSI, YURY - BIGGS, MARK - SALANNE, MATHIEU - SIMON, PATRICE - KANEKO, KATSUMI
0212	CHARGE-DISCHARGE CHARACTERISTICS OF GRAPHENE LIKE GRAPHITE PARTIALLY REDUCED BY LITHIUM METAL MATSUO, YOSHIAKI - SHUNYA, SHUNYA - INAMOTO, JUN-ICHI - MAEDA, KATSUMI - TAMURA, NORIYUKI
0226	SYNTHESIS OF CARBON FILAMENTS AND HIERARCHICAL COMPOSITES VIA NON-CATALYTIC DECOMPOSITION OF HALOGENATED HYDROCARBONS BAUMAN, YURII - MISHAKOV, ILYA - VEDYAGIN, ALEKSEY
0228	SYNTHESIS OF CARBON NANOTUBE FIBER USING HYDROCARBON WITH LOW DECOMPOSITION TEMPERATURE IN THE VAPOR PHASE LEE, SUNG-HYUN - KIM, HYE-RIM - LEE, HAEMIN - PARK, JUNBEOM - LEE, KUN-HONG - MIN KIM, SEUNG - LEE, JAEGEUN
0230	MODELLING PHYSICAL ADSORPTION ON FINITE OBJECTS USING 3 DIMENSIONAL NON-LOCAL DENSITY FUNCTIONAL THEORY 1: INITIAL STUDY OLIVIER, JAMES
0231	SELECTIVE CONTROL OF SURFACE HYDROPHILICITY DEPENDING ON PORE SIZE OF POROUS CARBON MATERIALS YU, YAO - NAKABAYASHI, KOJI - YOON, SEONG-HO - MIYAWAKI, JIN
0234	NOVEL ONE POT GREEN APPROACH FOR SYNTHESIS OF WELL EXFOLIATED MULTILAYER NITROGEN-DOPED GRAPHENE NANOSHEETS MOHAMED, MARWA - HAMAD, HESHAM - ELESSAWY, NOHA - CARRASCO-MARÍN, FRANCISCO
0237	ULTRAFAST HYDROTHERMAL ASSEMBLY OF NANOCARBON MICROFIBERS IN NEAR-CRITICAL WATER CHEN, YUAN - ZHAI, SHENGLI
0244	POROUS THERMOSET POLYMERS DERIVED BY HUMINS BIOREFINERY SIDE STREAM TOSI, PIERLUIGI - MIJA, ALICE - VAN KLINK, GERARD
0248	EXCELLENT ORR CATALYSTS BASED ON N-DOPED CARBON MATERIALS PREPARED THROUGH HEAT TREATMENT OF PANI AT HIGH TEMPERATURES QUILEZ-BERMEJO, JAVIER - MORALLÓN, EMILIA - CAZORLA-AMORÓS, DIEGO
0249	GLUCOSE-DERIVED CARBONS/CARBON NANOTUBE HYBRIDS FOR THE OXYGEN REDUCTION REACTION REY-RAAP, NATALIA - G. MORAIS, RAFAEL - FIGUEIREDO, JOSÉ LUÍS - R. PEREIRA, M. FERNANDO
0250	PREPARATION OF 3D-PRINTED GRAPHENE MONOLITHS AND THEIR APPLICATION AS WET PEROXIDE OXIDATION CATALYSTS FOR WASTEWATER TREATME QUINTANILLA, ASUNCION - CASAS, JOSE ANTONIO - MIRANZO, PILAR - OSENDI, MARIA ISABEL - BELMONTE, MANUEL
0251	TAILORING POROUS CARBON FIBRES FOR ENERGY STORAGE IN THE STABILISATION PROCESS WOLZ, DANIEL SEBASTIAN - RICHTER, BENJAMIN - BÖHM, ROBERT - JÄGER, HUBERT - RICHTER, MIRKO - TRAN, NGUYEN HOAI AN - CHERIF, CHOKRI - HÄRTEL, PAUL - BÖNKE, TOM - KASKEL, STEFAN
0261	IMPROVING ELECTRICAL CONDUCTIVITY AND ELASTICITY OF CARBON NANOTUBE COMPOSITES FOR FLEXIBLE AND STRETCHABLE ELECTRONICS HWANG, JI-YOUNG - LEE, SANG-HOON - SHIN, UEON SANG - CHOI, YOUNG CHUL
0264	CARBON NANODOTS FOR BIOLOGICAL APPLICATIONS RYBARCZYK, MARIA K. - GONTAREK, EMILIA - OLLIK, KAROLINA - LIEDER, MAREK
0265	VALORIZATION OF WASTED COFFEE POWDER FROM COFFEE-MACHINE AS ACTIVATED CARBONS BY SALT SHIELD PROCESS MESA-BRIBIAN, ELENA M - ALVAREZ-RODRIGUEZ, JESUS - MAROTO-VALIENTE, ANGEL
0266	SYNTHETIC PROCESS OF C@AL2O3 HYBRIDS WITH HYDROPHOBIC AND MAGNETIC PROPERTIES FOR ENVIRONMENTAL APPLICATIONS ORTEGA-MORENO, M JOSE - ALVAREZ-RODRIGUEZ, JESUS - MAROTO-VALIENTE, ANGEL
0268	SINGLE-WALLED CARBON NANOTUBES AS FILLERS IN NANOSTRUCTURED TITANIUM DIOXIDE ELECTRODES ANSÓN-CASAOS, ALEJANDRO - HERNÁNDEZ-FERRER, JAVIER - SANTIDRIÁN, ANA - GARCÍA-BORDEJÉ, ENRIQUE - BENITO, ANA - MASER, WOLFGANG
0275	ACTIVE CARBON MONOLITHS FROM BROWN COAL PARSA, MEHRDAD - QI, EMMA - CHAFFEE, ALAN L
0276	THE ACTIVE SITE, DOPING EFFECT AND REACTION PATHWAY OF THE DEHYDROGENATION OF ALKANES ON THE NANOSTRUCTURED CARBON CATALYST LI, BO
0279	SOLVENT-FREE SOFT-TEMPLATING SYNTHESIS OF MESOPOROUS CARBONS FOR ELECTROCHEMICAL CAPACITORS NISHIYAMA, NORIKAZU - YOSHIDA, NAO - HIROTA, YUICHIRO - UCHIDA, YOSHIAKI - ASADA, TOSHIHIRO - KOBAYASHI, NAOYA

Code	Title
0283	ENGINEERED GRAPHENE QUANTUM DOTS FOR CATALYSIS CHEN, PENG
0284	BACTERIA-ASSISTED SYNTHESIS OF CARBON -INORGANIC NANOPARTICLES COMPOSITES KOTARBA, ANDRZEJ - DUCH, JOANNA - GOLDA-CEPA, MONIKA - ZAPRZALSKA, ALICJA - OCHOŃSKA, DOROTA - BRZYCH-CZY-WLOCH, MONIKA - INDYKA, PAULINA - SOJKA, ZBIGNIEW
0285	ONE-DIMENSIONAL LINEAR CARBON CHAINS: FROM POLYYNE TO CARBYNE SHI, LEI - ROHRINGER, PHILIP - LAPIN, ZACHARY J - HEEG, SEBASTIAN - NOVOTNY, LUKAS - CAMBRE, SOFIE - WENSELEERS, WIM - WASSERROTH, SÖREN - REICH, STEPHANIE - WANKO, MARIUS - RUBIO, ANGEL - SUENAGA, KAZU - AYALA, PAOLA - KUZMANY, HANS - PICHLER, THOMAS
0287	BINDERLESS ORDERED MESOPOROUS CARBON MONOLITHS FROM LOW VALUE COAL TAR-DERIVED PRODUCTS CASTRO-MUÑIZ, ALBERTO - LORENZO-FIERRO, SARA - MARTÍNEZ-ALONSO, AMELIA - TASCÓN, JUAN MD - SUÁREZ-GARCÍA, FABIÁN - PAREDES, JUAN I
0288	GRAPHENE-POLY(E-CAPROLACTONE) MIXED-MATRIX FILMS MODULATE THE DIFFERENTIATION OF NEURAL PROGENITOR CELLS SANCHEZ-GONZALEZ, SANDRA - BIANCHI, FABIO - YE, HUA - URTIAGA, ANE - DIBAN, NAZELY
0291	REAL-TIME OBSERVATION OF CARBON BLACK PARTICLE IN AN ENVIRONMENTAL TRANSMISSION ELECTRON MICROSCOPE NASERI, ALI - SEDIAKO, ANTON - THOMSON, MURRAY
0295	IMPROVING THE ACTIVITY OF PD/C CATALYSTS USING N-DOPED SUPPORTS OBTAINED BY MODIFICATION OF COMMERCIAL ACTIVATED CARBONS RUIZ-GARCÍA, CRISTINA - HERAS, FRANCISCO - ALONSO-MORALES, NOELIA - CALVO, LUISA - GILARRANZ, MIGUEL A - RODRIGUEZ, JUAN J
0296	STUDY ON SEPARATING GRAPHITE FROM HTGR ELEMENTS BY ELECTROCHEMICAL METHOD WEN, MINGFEN - WANG, SHUWEI - CHEN, JING
0301	LIGNIN-BASED CARBON FIBRES: AN APPROACH TOWARDS NEW MARKET PERSPECTIVES FOLLMER, MARIE - JESTIN, SIMON - NERI, WILFRID - DERRE, ALAIN - BENTALEB, AHMED - MERCADER, CELIA - POULIN, PHILIPPE
0302	POROUS POLYANILINE/REDUCED GRAPHENE OXIDE COMPOSITE AS A BINDER-FREE ELECTRODE MATERIAL FOR HIGH-PERFORMANCE SUPERCAPACITORS MOYSEOWICZ, ADAM - GRYGLEWICZ, GRAŻYNA
0303	CARBON-MEDIATED SOLAR PHOTOCATALYSIS OF INDUSTRIAL WASTEWATER: ECOTOXICITY ASSESSMENT LOURENÇO, VÂNIA - ANIA, CONCHI - FONSECA, ISABEL
0304	LIGNIN-BASED CARBON NANOFIBERS MADE BY ELECTROSPINNING ROMAN, JULIEN - JESTIN, SIMON - MERCADER, CÉLIA - DERRE, ALAIN - POULIN, PHILIPPE
0305	1H-NMR STUDY ON THE THERMOPLASTIC PHENOMENON OF COAL WITH HIGH PERFORMANCE COKING ADDITIVE. KUMAGAI, HARUO - SAKAI, KOJI - OKUYAMA, NORIYUKI
0306	IMPACT OF WATER VAPOR ON BIOGAS SEPARATION BY MEANS OF ADSORPTION ON A BIOMASS BASED CARBON DURÁN, INÉS - ÁLVAREZ- GUTIÉRREZ, NOELIA - RUBIERA, FERNANDO - PEVIDA, COVADONGA
0308	MEASURING HEAT CAPACITY OF ACTIVATED CARBONS FOR CO2 CAPTURE QUEREJETA, NAUSIKA - GARCÍA, SUSANA - ÁLVAREZ-GUTIÉRREZ, NOELIA - RUBIERA, FERNANDO - PEVIDA, COVADONGA
0313	FE-CATALYZED GRAPHITIC CARBON MATERIALS FROM BIOMASS RESOURCES AS ANODES FOR LITHIUM ION BATTERIES GÓMEZ-MARTÍN, AURORA - RUTTERT, MIRCO - PLACKE, TOBIAS - MARTINEZ-FERNANDEZ, JULIAN - RAMÍREZ-RICO, JOAQUÍN
0318	PROPERTY EVALUATION OF CARBON FIBER/ABS COMPOSITES WITH DIFFERENT A, B, AND S COMPOSITIONS VIA LONG FIBER THERMOPLASTIC-BASED CHO, DONGHWAN - LEE, HEESOOK
0321	A SUPER-LONG CARBON NANOTUBE ARRAY FOR SIZE-BASED VIRUS CAPTURE AND DETECTION YEH, YIN-TING - TERRONES, MAURICIO
0323	CHRONOAMPEROMETRIC RESPONSE OF NANOPOROUS CARBONS UNDER UV-VIS IRRADIATION GOMIS BERENGUER, ALICIA - ANIA, CONCHI
0326	HIGHLY DISPERSED RUTHENIUM NANOPARTICLES ANCHORED ON CARBON NANOSPHERES AS HIGH EFFICIENT CATALYST FOR THE REMOVAL OF BPA SERRA-PÉREZ, ESTRELLA - ÁLVAREZ-TORRELLAS, SILVIA - DELGADO, JOSÉ ANTONIO - AGUEDA, VICENTE ISMAEL - OVEJERO, GABRIEL - GÓMEZ DE LA CRUZ, VICTORIA - GARCÍA, JUAN
0328	DIRECT SPECIATION SEPARATION OF OXYGEN FUNCTIONAL GROUPS BY NON-LINEAR THERMAL ANALYSIS DING, YUXIAO - GU, QINGQUING - SCHLÖGL, ROBERT - HEUMANN, SASKIA - DUENGEN, PASCAL

Code	Title
0331	IMPROVING CARBON/CARBON SUPERCAPACITORS CELL VOLTAGE IN AQUEOUS ELECTROLYTES WEISSMANN, MARTIN - BROUSSE, THIERRY - RAYMUNDO-PINERO, ENCARNACION
0346	MODELING AND INVESTIGATION OF CROSS-LINKING EFFECT ON MECHANICAL PROPERTIES OF CNT YARNS CHO, YOUNG SHIK - JUNG, YEONSU - PARK, CHONG RAE
0348	NANO-GRAPHITE AND RATE PERFORMANCE OF LITHIUM INTERCALATION OGUMI, ZEMPACHI - ABE, TAKESHI
0349	PREPARING CARBON FIBER FROM NEEDLE COKE LYU, YONGGEN - WU, ZHENGNAN - YANG, CHANGLING
0355	IMPROVED PERFORMANCE OF AMORPHOUS CARBON MEMBRANES BY RESTRUCTURING DURING HIGH PRESSURE WATER DESALINATION ORTIZ-MEDINA, JOSUE - KITANO, HIROKI - ITO, ISAMU - CRUZ-SILVA, RODOLFO - MORELOS-GOMEZ, AARON - TAKEUCHI, KENJI - TERRONES, MAURICIO - HAYASHI, TAKUYA - ENDO, MORINOBU
0357	CHEMICAL VAPOR DEPOSITION GROWTH OF NANOCRYSTALLINE GRAPHENE FILMS AND ITS MECHANICAL PROPER-TIES GAO, LIBO
0358	ATOMISTIC SIMULATIONS OF GLASSY CARBON UNDER HIGH PRESSURES DE TOMAS, CARLA - SUAREZ-MARTINEZ, IRENE - SHIELL, THOMAS B. - MCCULLOCH, DOUGAL G. - MCKENZIE, DAVID R. - FIELD, MAT-THEW R. - HABERL, BIANCA - BOEHLER, REINHARD - COOK, BRENTON A - MARKS, NIGEL A. - BRADBY, JODIE E.
0359	HIERARCHICAL CELLULOSE-DERIVED CARBON COMPOSITE ELECTRODES FOR SUSTAINABLE SUPERCAPACITORS KUZMENKO, VOLODYMYR - HAQUE, MAZHARUL - LI, QI - NARGES, NASERI - SMITH, ANDERSON D. - LUNDGREN, PER - GATENHOLM, PAUL - AJI, MATHEW - ENOKSSON, PETER
0363	EFFECT OF NITROGEN ON THE PERFORMANCE OF SENSOR BASED ON N-DOPED REDUCED GRAPHENE OXIDES FOR THE DETERMINATION OF EPINEPHRINE WIENCH, PIOTR - GONZÁLEZ, ZORAIDA - MENÉNDEZ, ROSA - GRYGLEWICZ, GRAŻYNA
0366	REASSESSING COAL'S SHARE IN ENERGY MIX: A GLOBAL PERSPECTIVE AND SUSTAINABLE WASTE COAL STRATEGY CLEMENTE-JUL, CARMEN - GUERRERO, FRANCISCO
0374	UNDERSTANDING THE GRAPHENE FOAM ELABORATION PREPARED FROM A SOLVOTHERMAL-BASED PROCESS: ROLE OF THE SODIUM COMPOUNDS SPEYER, LUCIE - FONTANA, SEBASTIEN - CAHEN, SEBASTIEN - HEROLD, CLAIRE
0376	CELLULOSE-BASED CARBON FIBERS: INCREASING TENSILE STRENGTH AND CARBON YIELD UNTERWEGER, CHRISTOPH - HINTERREITER, ANDREAS - STIFTER, DAVID - FÜRST, CHRISTIAN
0380	BIOMIMETIC STRONTIUM-SUBSTITUTED HYDROXYAPATITE DEPOSITED ON CARBON FIBER CLOTH FOR BONE REGEN-ERATION OLIVIER, FLORIAN - PICARD, QUENTIN - DELPEUX, SANDRINE - FAYON, FRANCK - VINCENT, SAROU-KANIAN - CHANCOLON, JEROME - FABIENNE, WARMONT - ROCHET, NATHALIE - BONNAMY, SYLVIE
0381	POROUS CARBON TEMPLATES FROM THERMOSET-BASED WOOD POLYMER COMPOSITES AS PRECURSOR FOR SILI-CON CARBIDE CERAMICS UNTERWEGER, CHRISTOPH - PLANK, BERNHARD - SENCK, SASCHA - MIHALIC, MATTHIAS - FÜRST, CHRISTIAN
0391	MICROPOROUS ELECTROSPUN LIGNIN-BASED FIBERS AS BINDERLESS SUPERCAPACITOR ELECTRODES GARCÍA-MATEOS, FRANCISCO JOSÉ - RUIZ-ROSAS, RAMIRO - MORALLON, EMILIA - CAZORLA-AMORÓS, DIEGO - RODRÍGUEZ-MIRA-SOL, JOSÉ - CORDERO, TOMÁS
0392	ONE-POT SINTHESYS OF LIGNIN-BASED ELECTROSPUN OXYGEN REDUCTION REACTION ELECTROCATALYSTS GARCÍA-MATEOS, FRANCISCO JOSÉ - RUIZ-ROSAS, RAMIRO - CAZORLA-AMORÓS, DIEGO - MORALLÓN, EMILIA - RODRÍGUEZ-MIRA-SOL, JOSÉ - CORDERO, TOMÁS
0393	FIXED-BED ADSORPTION OF BISPHENOL A ONTO CARBON ADSORBENTS: ADSORPTION/DESORPTION CYCLES HERNÁNDEZ-ABREU, ANA BELÉN - GOFFIN-MEDINA, SHAROL - NIETO-NARANJO, PATRICIA - ÁLVAREZ-TORRELLAS, SILVIA - ÁGUE-DA, VICENTE ISMAEL - DELGADO, JOSE ANTONIO - OVEJERO, GABRIEL - GARCÍA, JUAN
0395	ANTICORROSIVE EFFECTS OF CARBON/SEMICONDUCTOR COATINGS ON METALS QUEZADA RENTERIA, JAVIER ALAN - GOMIS BERENGUER, ALICIA - CHÁZARO RUIZ, LUIS FELIPE - RANGEL MENDEZ, JOSE RENE - ANIA, CONCHI
0403	PHOTOCATALYTIC DEGRADATION OF PERFLUOROCARBOXYLIC ACIDS BY TIO2-RGO COMPOSITE DIBAN, NAZELY - RIBAO, PAULA - RIVERO, MARIA J. - ORTIZ, INMACULADA - URTIAGA, ANE - GOMEZ, BEATRIZ
0404	COMPARATIVE MICROSTRUCTURAL ANALYSIS OF NON-GRAPHITIC CARBONS BASED ON WIDE-ANGLE X-RAY (WAXS) AND NEUTRON (WANS) SCATTERING PFAFF, TORBEN - BADACZEWSKI, FELIX - SMARSLY, BERND M.
0410	MOLECULAR BEHAVIOR IN GRAPHENE NANOWINDOWS DOMAINS VALLEJOS-BURGOS, FERNANDO - KANEKO, KATSUMI

Code	Title
0411	COMBUSTION KINETICS OF CHAR DERIVED FROM WASTE TIRE PYROLYSIS – A COMPARISON OF DIFFERENT STRUCTURAL MODELS MARTINEZ, JUAN DANIEL - ARENAS, CINDY NATALIA - MURILLO, RAMON - NAVARRO, MARIA VICTORIA - GARCIA, TOMAS
0412	CHARGE-TRANSFER PROCESSES BETWEEN GRAPHENE AND SIMPLE MOLECULES OR CARBON NANODOTS MESSINA, FABRIZIO - BUSCARINO, GIANPIERO - MESSINA, FABRIZIO - SCIORTINO, ALICE - CANNAS, MARCO - GELARDI, FRANCO MARIO - GIANNAZZO, FILIPPO - SCHILIRÒ, EMANUELA - AGNELLO, SIMONPIETRO
0415	EFFECT OF OXYGEN IMPURITY ON VOC DESORPTION FROM ACTIVATED CARBON HASHEMI, SEYED MOJTABA - HASHISHO, ZAHER - PHILLIPS, JOHN H - ANDERSON, JAMES - NICHOLS, MARK
0422	STRUCTURE AND STABILITY OF INTERCALATED MULTILAYER GRAPHENE MATSUMOTO, RIKA - AKUZAWA, NOBORU
0423	CARBON FIBERS FROM AROMATIC POLYMER PRECURSORS: PRECURSOR DESIGN FOR LOW CO2 EMISSION PRODUCTS HATORI, HIROAKI - SAKAMOTO, KAITO - KAWAJIRI, KOTARO - TAHARA, KIYOTAKA
0424	STRUCTURAL CHANGE OF GRAPHITE ULTRATHIN FILMS FROM SPIN-COATING POLYMERS SONEDA, YASUSHI - YOSHIZAWA, NORIKO - KODAMA, MASAYA
0425	HIGH-DENSITY SINGLE IRON ATOM-DISPERSED N-DOPED POROUS CARBON FOR EFFICIENT OXYGEN REDUCTION UNDER ACIDIC MEDIUM HOU, PENGXIANG - LI, JINCHENG - LIU, CHANG - CHENG, HUIMING
0426	EXPERIMENTAL AND THEORETICAL STUDY OF AGEING MECHANISM OF ACTIVATED CARBON BASED SUPERCAPACITOR AND SOME COUNTERMEASURES HUANG, YILUN - GONG, QIANMING
0433	EFFECT OF MWCNT COMPOSITION ON THE PERMSELECTIVITY OF PES-MWCNT MIXED MATRIX MEMBRANES FOR GAS SEPARATION LINCK, NICHOLAS - WEISENBERGER, MATTHEW - KALIKA, DOUGLASS
0435	GREEN SYNTHESIS OF ULTRALIGHT POROUS GRAPHENE COCOON FOR ENHANCED MICROWAVE ABSORBING PERFORMANCE WANG, SHANSHAN - JIAO, QINGZE - XUE, HAOLIANG - GAO, MEIMEI - FENG, CAIHONG - SHI, DAXIN - ZHAO, YUN
0436	HIGHLY DISPERSED CO3S4 WITHIN GRAPHITIC CARBON POLYHEDRONS@CARBON NANOTUBE SPONGE FOR ULTRALONG LIFE LITHIUM-SULFUR BATTERIES ZHANG, HUI
0437	OPTIMUM POROUS CARBON STRUCTURE FOR HIGH CAPACITY EDLC ENHANCED BY SPECIFIC ION-DESOLVATION URITA, CHIHARU - URITA, KOKI - FUJITA, KOSUKE - HORIO, KEIJI - YOSHIDA, MASAYUKI - MORIGUCHI, ISAMU
0439	DISLOCATION MOTION AND DYNAMICS IN BILAYER GRAPHENE EXTENDED TO RADIATION DAMAGE OF NUCLEAR GRAPHITE IN REACTORS MOURATIDIS, PAVLOS - JOLLEY, KENNY - HEGGIE, MALCOLM I - MAERZ, BEN - WU, HOUZHENG - SMITH, ROGER - RAYSON, MARK - BRIDDON, PATRICK R
0440	SCALABLE FABRICATION OF POROUS CNTS BASED MICROSPHERES AND THEIR APPLICATION IN EXTRACTION OF CHINESE TRADITIONAL MEDICINE GONG, QIANMING - BAI, JUNFEI - LI, XI - HUANG, YILUN - GAN, JIANNING - ZHAO, MING - ZHUANG, DAMING - LIANG, JI
0442	INVESTIGATING THE EFFECT OF LONG RANGE ELECTROSTATIC INTERACTIONS OF CURVED CARBON MATERIALS IN ION-INDUCED NUCLEATION BOWAL, KIMBERLY - MARTIN, JACOB - KRAFT, MARKUS
0443	PHOTOCATALYTIC OXIDATION OF PROPANE WITH TIO2 AND TIO2-C PHOTOCATALYSTS CANO-CASANOVA, LAURA - LILLO-RÓDENAS, MARÍA ÁNGELES - ROMÁN-MARTÍNEZ, MARÍA DEL CARMEN - MEI, BASTIAN - MUL, GUIDO
0445	ADSORPTION OF DICLOFENAC FROM AQUEOUS SOLUTION ONTO CARBON XEROGELS. EFFECT OF TEXTURAL PROPERTIES AND PRESENCE OF BACTERIA MORAL-RODRIGUEZ, ADRIANA ISABEL - LEYVA-RAMOS, ROBERTO - CARRASCO-MARIN, FRANCISCO - OCAMPO PEREZ, RAUL - BAUTISTA-TOLEDO, MARIA ISIDORA
0446	NEW PARADIGMS FOR ANALYZING X-RAY DIFFRACTION SPECTRA OF GRAPHENE-BASED CARBONS PUECH, PASCAL - AGNIESZKA, DABROWSKA - NICOLAS, RATEL-RAMOND - GERARD, VIGNOLES - MARC, MONTHIOUX
0447	KEROGEN FLEXIBILITY, A KEY TO UNDERSTAND HYDROCARBON EXPULSION FROM SHALE RESERVOIRS ? OBLIGER, AMAEL - VALDENAIRE, PIERRE-LOUIS - CAPIT, NICOLAS - ULM, FRANZ - PELLENQ, ROLAND - LEYSSALE, JEAN-MARC
0453	HARD CARBON DERIVED FROM BLENDING POLYMER INTERPENETRATING NETWORKS FOR SODIUM-ION BATTERIES ANODE ZHANG, WENFENG - ZHANG, HUIMIN - ZHANG, HAO - MING, HAI - CAO, GAOPING

Oral Index

Code	Title
0456	THERMODYNAMICS AND KINETICS OF GRAPHENE AND H2O/CO2 REACTIONS: INSIGHTS OF MULTIPLE-SITE EDGE EFFECTS. OYARZÚN-ARAVENA, ANDREA M. - SALGADO-CASANOVA, ADOLFO JA - RADOVIC, LJUBISA R - GARCÍA-CARMONA, XIMENA
0458	AQUEOUS MERCURY REMOVAL FROM ACTUAL SFGD WASTEWATER OF A COAL-FIRED POWER PLANT USING SULFURIZED ACTIVATED CARBON CHEN, YUN-HSIN - HSU, CHE-JUNG - HSI, HSING-CHENG
0459	CO2 ADSORPTION BEHAVIOR OF POLLEN-DERIVED POROUS CARBON CHOI, SEUNG WAN - TANG, JIALIANG - POL, VILAS - LEE, KI BONG
0466	NICKEL-CARBON NANOPARTICLES: SIZE DEPENDENT PHASE DIAGRAMS AND INTERACTION WITH GRAPHENIC LAYERS YANN, MAGNIN - ROLAND, PELLENQ - CHRISTOPHE, BICHARA - HAKIM, AMARA
0469	EFFECT OF GRAPHENE OXIDE SYNTHESIS METHOD ON PERFORMANCE OF POLYSULFONE-GRAPHENE OXIDE MIXED MATRIX MEMBRANE SALI, SAFAE - MACKEY, HAMISH - ABDALA, AHMED
0470	CARBON NANOGRAPES: A NEW COMPOSITE MATERIAL FOR USE AS ELECTRODE ON SUPERCAPACITORS ARENAS ESTEBAN, DANIEL - ÁVILA BRANDE, DAVID - OTERO DÍAZ, L. CARLOS
0471	MICROWAVE HYDROGEN COLD PLASMA AS REDUCING AGENT FOR GRAPHENE OXIDE ABDELKADER FERNÁNDEZ, VÍCTO K. ABDELKADER FERNÁNDEZ - DOMINGO GARCIA, MARIA - LÓPEZ GARZÓN, F JAVIER - MELGUIZO, MANUEL - LÓPEZ DE LA CALLE, Mª DOLORES - PÉREZ MENDOZA, MANUEL J
0473	ULTRALIGHT, FLEXIBLE, AND FREE-STANDING CARBON MICROTUBE MEMBRANE SYNTHESIZED BY SI3N4 NANOWIRE-TEMPLATED CHEMICAL VAPOR DEPO YIN, XUEMIN - LI, HEJUN - LU, JINHUA
0474	ELECTROCHEMICAL PREPARATION OF HYDROXIDE-ION CONTAINING GRAPHITE INTERCALATION COMPOUNDS MIYAHARA, YUTO - FUKUTSUKA, TOMOKAZU - ABE, TAKESHI
0476	ADSORPTION GAS TERMINAL ON THE BASIS OF NEW FUNCTIONAL CARBON NANOPOROUS MATERIAL FOR NATURAL GAS STORAGE AND TRANSPORTATION MEN'SHCHIKOV, ILYA - FOMKIN, ANATOLY - SHKOLIN, ANDREY - AKSYUTIN, OLEG - ISHKOV, ALEKSANDR - LADYGIN, KONSTANTIN - TSIVADZE, ASLAN
0479	BORON-DOPED GRAPHENE QUANTUM STRUCTURE@MOS2 NANOHYBRIDS AS HIGH PERFORMANCE ANODE FOR LITHIUM ION BATTERY DOONG, RUEY-AN - BINDUMADHAVAN, KARTICK
0482	THE ROLE OF SOLVENT INTERACTIONS AND CHARGE SEPARATION IN THE FUNDAMENTAL PHOTOCYCLE OF CARBON NANODOTS MESSINA, FABRIZIO - SCIORTINO, ALICE - SCIORTINO, LUISA - GAZZETTO, MICHELA - CANNIZZO, ANDREA - CANNAS, MARCO
0486	A FAMILY OF FLUORESCENT CARBON DOTS WITH HIGHLY HOMOGENEOUS STRUCTURAL PROPERTIES AND EXCEPTIONAL OPTICAL PERFORMANCE SCIORTINO, ALICE - MAURO, NICOLÒ - BUSCARINO, GIANPIERO - SCIORTINO, LUISA - POPESCU, RADIAN - SCHNEIDER, REINHARD - GIAMMONA, GAETANO - GERTHSEN, DAGMAR - CANNAS, MARCO - MESSINA, FABRIZIO
0488	ACTIVATED CARBON FOR GOLD ADSORPTION USING COLOMBIAN COAL AS RAW MATTER MARTÍNEZ MENDOZA, KAREN LORENA - BARRAZA BURGOS, JUAN MANUEL - MARRIAGA, NILSON DE JESÚS - MACHUCA MARTINEZ, FIDERMAN
0498	HIGH–PERFORMANCE OF GAS HYDRATES CONFINED IN NANOPOROUS SOLIDS FOR CH4 AND CO2 STORAGE CUADRADO-COLLADOS, CARLOS - MARTINEZ-ESCANDELL, MANUEL - CASCO, MIRIAN - SILVESTRE, JOAQUIN
0500	NUMERICAL SIMULATION OF FLUID DYNAMICS IN POROUS CARBON NANOMATERIALS BY ELECTROWETTING MA, RONG - YAO, WEI - LU, XIAOCHEN - ZHANG, GUANG
0501	RAMAN STATISTICAL ANALYSIS OF CVD MONOLAYER GRAPHENE. COMPARISON WITH LITERATURE KATHARINA, GRATZER - CEDRIC, PARDANAUD - ALEXANDRE, MERLEN - JOSEPHUS GERARDUS, BUIJNSTERS - OLIVIER, CHUZEL - LIONEL, PATRONE - SYLVAIN, CLAIR - PASCALE, ROUBIN - JEAN-LUC, PARRAIN
0502	IN SITU REINFORCEMENT OF METALS WITH GRAPHENE: HIGH-TEMPERATURE INTERACTIONS BETWEEN SIC AND METALS (IR, RH, PT, PD) NARCISO, JAVIER - ORTEGA, ADRIAN - CAMARANO, ANTONIO - CACCIA, MARIO - GIURANNO, DONATELLA - JUAN, JERONIMO
0510	FARADAIC BEHAVIOR OF QUINONE-GRAFTED CARBONS IN PROTIC IONIC LIQUID ELECTROLYTE GÓRSKA, BARBARA - BABUCHOWSKA, PAULINA - BÉGUIN, FRANÇOIS
0513	ELECTROCHEMICAL OVEROXIDATION OF GRAPHITE IN MIXTURE COMPOSED OF SULFURIC AND NITRIC ACID GURZEDA, BARTOSZ - BACHAR, AGNIESZKA - KRAWCZYK, PIOTR
0515	SOLUTION SPUN MNO2 DOPED CARBON NANOTUBE FIBERS FOR SEWABLE SUPERCAPACITORS TAYLOR, LAUREN - PASQUALI, MATTEO

Code	Title
0516	CRI PREDICTIVE STUDIES OF COKE BY DRIFTS ANALYSIS OF COLOMBIAN COALS ARIZA CUBILLOS, EDNA - CADENA SÁNCHEZ, ARIEL - ROMERO MALAGÓN, EDUARD - AGÁMEZ-PERTUZ, YAZMIN - DÍAZ VELÁSQUEZ, JOSÉ
0519	EVALUATION OF DURABLE FUEL CELL ELECTRODES USING THE PLATINUM- LOADED GRAPHENE. TOYODA, MASAHIRO - NAGATA, TAKEFUMI - SUTAN, KOUICHI - ARAI, YASUHIKO - KINUMOTO, TARO - TSUMURA, TOMOKI - MURA-MATSU, KAZUO
0522	CATALYTIC ACTIVITY OF CARBON MATERIALS WITH DIFFERENT STRUCTURES AND BY-PRODUCED CARBONS FOR HYDROGEN PRODUCTION SUDA, YOSHIYUKI - NISHII, HARUKI - MIYAMOTO, DAI - HAMAGUCHI, HIROAKI - SUZUKI, MASASHI - HARIGAI, TORU - TANIMOTO, TSUYOSHI - TAKIKAWA, HIROFUMI - UMEDA, YOSHITO
0525	HYDROGEN STORAGE IN HIGH-SURFACE AREA CARBON PELLETS OUZZINE, MOHAMMED - OHWADA, MAO - NISHIHARA, HIROTOMO - KYOTANI, TAKASHI
0526	ANALYSIS OF STACKING ORDER OF GRAPHENES IN TURBOSTRATIC GRAPHITE BY MOIRÉ INTERFEROMETRY OF TEM IMAGE FUJIMOTO, HIROYUKI - HISA, MASAAKI - AZUMA, TAKAYUKI - MORITA, KOUICHI
0531	EFFECTS OF PREHEATING ON ELECTRONIC TRANSPORT OF DETONATION NANODIAMONDS PINA-SALAZAR, ELDA-ZORAIDA - HAYASHI, TAKUYA - TOSHIO, SAKAI - OSAWA, EIJI - KANEKO, KATSUMI
0538	SYNTHESIS OF HIERARCHICALLY STRUCTURED CARBON-CARBON AND CARBON-MINERAL NANOCOMPOSITES FOR VARIOUS APPLICATIONS VEDYAGIN, ALEKSEY - MISHAKOV, ILYA - YURII, BAUMAN - DENIS, KORNEEV - ALEKSEY, VEDYAGIN
0539	EFFECTIVE REDUCTION OF GRAPHENE OXIDE BY A HYBRID MICROWAVE HEATING METHOD TANG, SHAN - JIN, SHUANGLING - LU, WANGZHAO - HU, ZHEN - YANG, SHUO - WANG, JIANGCAN - ZHANG, RUI - LIU, YAN - JIN, MINGLIN
0542	CONTROLLED SYNTHESIS OF COTTON FABRIC/NICO2O4 MATERIAL FOR LITHIUM-ION BATTERY XI, YUKUN - LIU, YAN - ZHANG, DENGKE - JIN, SHUANGLING - ZHANG, RUI - JIN, MINGLIN
0550	DEFECT ANALYSIS FOR CARBON FIBERS BASED ON STRENGTH REDUCTION IN LIQUIDS SHIOYA, MASATOSHI - MASAYOSHI, IDO - KAJIKAWA, TAKASHI - TAKAHASHI, KUNIAKI - SUGIMOTO, YOSHIKI
0552	DIMINISHING THE BACK DIFFUSION OF FE CATALYST PARTICLES FOR SYNTHESIZING LONG CNT FORESTS BY USING FE PRE-IMPLANTED WAFERS LEE, CHEOL-HUN - LEE, JAEGEUN - LEE, KUN-HONG
0554	NOVEL TECHNIQUE FOR MICROSTRUCTURE QUANTIFICATION OF CARBON COMPOSITES BY STEAM GASIFICATION HISA, MASAAKI - YANO, TOYO - AZUMA, TAKAYUKI
0556	FAR-FIELD ELECTROMAGNETIC CHARACTERIZATION OF PREPREG FILM BASED RADAR ABSORBING COMPOSITE CHOI, JAE RYUNG - JUNG, BYUNG MUN - CHO, SEUNGCHAN - PARK, BYEONGJIN - LEE, SANG BOK
0558	GRAPHENE OXIDE MEMBRANES FOR FOOD FILTRATION MORELOS-GOMEZ, AARON - CRUZ-SILVA, RODOLFO - ORTIZ-MEDINA, JOSUE - ARAKI, TAKUMI - TEJIMA, SYOGO - TAKEUCHI, KENJI - TERRONES, MAURICIO - ENDO, MORINOBU
0560	FORMATION OF THE METASTABLE SOLID-PHASE INTERMEDIATES DURING INTERACTION OF HALOGENATED COM-POUNDS WITH BULK NI-CR ALLOY KENZHIN, ROMAN - BAUMAN, YURI - VOLODIN, ALEXANDER - MISHAKOV, ILYA - VEDYAGIN, ALEKSEY
0561	LOCAL NANO-FUNCTIONALISATION OF GRAPHENE VIA CATALYTIC SCANNING PROBE NANOLITHOGRAPHY EVALUATED BY RAMAN SPECTROSCOPY GRATZER, KATHARINA - PARDANAUD, CÉDRIC - MERLEN, ALEXANDRE - CLAIR, SYLVAIN - BALABAN, SILVIU T. - CHUZEL, OLIVIER - PARRAIN, JEAN-LUC - ROUBIN, PASCALE
0563	STRUCTURAL MODIFICATION OF LOW MOLECULAR FRACTION OF FCC-DO BY USING MICROWAVE AND ELCTRON BEAM IRRADIATIONS IM, KYU KWAN - YANG, KAP SEUNG - KIM, DOO WON - KIM, YOONG AHM
0565	MECHANICAL PROPERTIES OF CARBIDE-DERIVED CARBONS DE TOMAS, CARLA - SUAREZ-MARTINEZ, IRENE - MARKS, NIGE
0568	EFFECT OF SURFACE MODIFICATION FOR SIZED CARBON FIBERS ON INTERLAMINAR SHEAR STRENGTH OF POLYAMIDE COMPOS-ITES YAO, LILI - LU, YONGGEN - ZHANG, WEISU - SUN, ZHULIN - SUN, YING - LIU, LIGUO - WU, ZHENGNAN
0570	ONE-POT SYNTHESIS OF SUPERACTIVATED CARBONS FOR HIGH POWER AQUEOUS SUPERCAPACITORS CARRIAZO, DANIEL - MYSYK, ROMAN - GOIKOLEA, EIDER - CARRIAZO, DANIEL
0573	THE RELATIONSHIP BETWEEN THE WETTABILITY OF COAL TAR PITCH AND THE THICKNESS OF THE BINDER LAYER IN GREEN ELECTRODE MATERIAL WOJTASZEK, MAŁGORZATA - TOPOLNICKA, TERESA

Oral Index

Code	Title
0578	PHOTOASSISTED PREPARATION OF METAL-CONTAINING NANOPOROUS POLYMERIC CARBONS C. FERNÁNDEZ DE CÓRDOBA, MARIA - ZAIER, MOHAMED - BALAN, LAVINIA - O. ANIA, CONCHI
0579	STUDY ON CHANGE OF SURFACE NATURE OF MULTI-WALLED CARBON NANOTUBES BY ACID TREATMENT FOR IMPROVING DISPERSIBILITY SHIN, MINHO - LIM, HYEONG JUN - PARK, CHONG RAE
0581	CARBON MICROTUBES DERIVED FROM ROLLED-UP CHITOSAN FILMS: THE DESIGN AND THE USE AS JOULE-HEATED MICROREACTORS LUCHNIKOV, VALERIY - SAITO, YUKIE - YAMADA, HAJIME - BEDA, ADRIAN - GHIMBEU, CAMELIA
0582	ONE-POT SELECTIVE CELLULOSE CONVERSION INTO SORBITOL USING A TANDEM CARBON CATALYST AZAR*, FATIMA-ZAHRA - LILLO RÓDENAS, MARIA ÁNGELES - ROMÁN-MARTÍNEZ, MARI CARMEN
0585	PREPARATION AND CHARACTERIZATION OF CARBON ALLOY CATALYSTS DERIVED FROM CHICKEN EGG FOR HYDROGEN EVOLUTION REACTION TAKIGAMI, MACHIKO - YOSHIKAWA, RISA - ISHII, TAKAFUMI - IMASHIRO, YASUO - OZAKI, JUN-ICHI
0588	ION CONDUCTING HYBRID ELECTROLYTE USING FUNCTIONALIZED SILICA PARTICLE FOR LI-ION BATTERY JUNG, BYUNG MUN - CHOI, U HYEOK - LEE, SANG-BOK
0589	HIGH SULFUR LOADING GRAPHENE COMPOSITES AS CATHODES FOR LI-S BATTERIES GÓMEZ URBANO, JUAN LUIS - GÓMEZ CÁMER, JUAN LUIS - DÍEZ, NOEL - BOTAS, CRISTINA - RODRIGUEZ, LIDE M. - ROJO, TEÓFILO - CARRIAZO, DANIEL
0594	CARBON NANOSPHERES AS MODIFYING AGENT OF SHALLOW FIELDS FOR ENHANCED PROCESS OF CARBON CAPTURE AND STORAGE-ECCS RODRIGUEZ-ACEVEDO, ELIZABETH - CORTES-CORREA, FARID - CARRASCO-MARIN, FRANCISCO - PEREZ-CADENAS, AGUSTIN
0599	EVOLUTION OF POROUS STRUCTURE AND THERMAL PROPERTIES OF EXPANDABLE AND EXPANDED GRAPHITE RESULTED BY GRAPHITE OXIDATION SHORNIKOVA, OLGA - KAMAEV, ALEXEY - TIKHONOV, NICKOLAI - FILIMONOV, STANISLAV - MALAKHO, ARTEM - AVDEEV, VIKTOR
0602	CATALYTIC ACTIVITY OF CARBON CATALYSTS WITH CORRUGATED GRAPHITIC LAYERS FOR OXYGEN REDUCTION REACTION IN ACIDIC CONDITIONS OZAKI, JUN-ICHI - ISHII, TAKAFUMI - KOBAYASHI, RIEKO - TAKIGAMI, MACHIKO - IMASHIRO, YASUO
0603	MOLECULAR UNDERSTANDING OF THE ACTIVE SITES ON BN-DOPED CARBON CATALYSTS FOR OXYGEN REDUCTION REACTION ISHII, TAKAFUMI - KOBAYASHI, RIEKO - TAKIGAMI, MACHIKO - IMASHIRO, YASUO - OZAKI, JUN-ICHI
0606	EFFECT OF LAYERED STRUCTURE OF GRAPHENE OXIDE MEMBRANE ON PERMEATION EFFICIENCY OF WATER ZHENG, XIAOHUI - DING, SONGTAO - WANG, RAN - XU, BIN - YAN, XIAOSHAN
0607	HIGH-PERFORMANCE UV PHOTODETECTOR BASED ON LENGTH-CONTROLLED ZNO NANOROD LEE, KANG EUN - KIM, NARI - PARK, TEAHOON - OH, YOUNGSEOK - YOO, JUNG-KEUN - UM, MOON-KWANG
0619	SYNTHESIS OF FE-N-DOPED CARBON NANOWALLS FOR OXYGEN REDUCTION REACTION YOSHIE, RYO - TANIGUCHI, MOEKA - AKIKUBO, KAZUMA - TATENO, AKIRA - KAWAGUCHI, NORIHITO - TACHIBANA, MASARU
0621	POLYMERIZATION OF SOLVATED C60 CRYSTALS UNDER PRESSURE GONNOKAMI, HIROMICHI - MURATA, HIDENOBU - TACHIBANA, MASARU
0626	ACTIVATED CARBON SUPPORTING BIOTRICKLING FILTERS FOR SILOXANE REMOVAL IN BIOGAS UPGRADING SANTOS-CLOTAS, ERIC - CABRERA-CODONY, ALBA - MARTÍN, MARIA J. - GICH, FREDERIC - BOADA, ELLANA - COMAS, JOAQUIM
0628	SYNTHESIS OF SULFONATED MULTIWALLED CARBON NANOTUBE USING SUPERCRITICAL CARBON DIOXIDE DURIAN, ALVA - MACAWILE, MARIA CRISTINA - GALLO, JUREX - AURESENIA, JOSEPH
0632	SYNTHESIS OF ORGANOSILANE – MWCNT CATALYST USING SUPERCRITICAL CARBON DIOXIDE FOR BIODIESEL PRODUCTION MACAWILE, MARIA CRISTINA - DURIAN, ALVA - GALLO, JUREX - RAZON, LUIS - TAN, RAYMOND - AURESENIA, JOSEPH
0637	CHARGE-TRANSFER COMPLEXES OF 2D SHEETS OF GRAPHENE OXIDE AND NANOPARTICLES OF POLY(3-HEXYL THIOPHENE) MASER, WOLFGANG - ISTIF, EMIN - HERNÁNDEZ-FERRER, JAVIER - COLCHERO, JAIME - PALACIOS-LIDÓN, ELISA - KLYMOV, OLEKSII - GARRÓ, NÚRIA - CROS, ANA - BENITO, ANA M.
0647	CHARACTERIZATION OF THE ATOMIC STRUCTURE OF GLASSY CARBON FOAMS JURKIEWICZ, KAROLINA - LISZKA, BARBARA - DUBER, STANISLAW - BURIAN, ANDRZEJ
0648	HIGHLY DURABLE ELECTROCATALYSTS WITH THERMALLY TREATED CARBON NANOFIBER FOR PEMFCS JI, YUNSEONG - HWANG, HOJUNG - LEE, CHANMIN - KWON, OH CHAN - JEON, OKSUNG - KIM, JEONG PIL - SHUL, YONG-GUN - JEON, YUKWON

Code	Title
0649	THE FABRICATION OF CONTROLLED THICKNESS GRAPHENE OXIDE FILMS BY MEANS OF DIP-ASSISTED LAYER BY LAYER ASSEMBLY KANDJOU, VEPIKA - HERNAEZ, MIGUEL - ACEVEDO-MUNOZ, BEATRIZ - MAYES, ANDREW - MELENDI-ESPINA, SONIA - PEREZ MAS, ANA MATILDE
0651	CARBON NANOTUBES EFFECT ON THE MORPHOLOGY AND TRANSPORT PROPERTIES OF AROMATIC POLYAMIDE REVERSE OSMOSIS MEMBRANES CRUZ-SILVA, RODOLFO - IZU, KAZUO - ORTIZ-MEDINA, JOSUE - MORELOS-GOMEZ, AARON - TAKIZAWA, YOSHIHIRO - OBATA, MICHIKO - MURAMATSU, HIROYUKI - TAKEUCHI, KENJI - HAYASHI, TAKUYA - TERRONES, MAURICIO - ENDO, MORINOBU
0654	PD CATALYSTS SUPPORTED ON ACTIVATED CARBONS PREPARED FROM LIGNIN FOR THE VALORIZATION OF CHLOROFORM BY HYDRODECHLORINATION FERNANDEZ-RUIZ, CARLOS - BEDIA, JORGE - RODRÍGUEZ, JUAN JOSÉ - GÓMEZ-SAINERO, LUISA MARÍA
0658	IN SITU X-RAY DIFFRACTION STUDIES WITHIN AN OPERATING ELECTROCHEMICAL CELL: FROM STRUCTURAL CONTROLLED MODIFICATIONS TO ION'S SANTOS, CLEIS - VILATELA, JUAN JOSÉ - MARCILLA, REBECA - PENDASHTED, AFSHIN - SENOKOS, EVGENY
0659	DOPED CARBON NANOTUBES PREPARED BY A SOLVENT-FREE METHODOLOGY FOR CWAO P. ROCHA, RAQUEL - G.P. SOARES, OLÍVIA SALOMÉ - J.M. ÓRFÃO, JOSÉ - F. R. PEREIRA, MANUEL - L. FIGUEIREDO, JOSÉ
0661	ONE-POT, FULLY SCALABLE AND LOW COST METHOD FOR CARBONACEOUS NANOSTRUCTURES FUNCTIONALIZATION ŻELECHOWSKA, KAMILA - NADOLSKA, MAŁGORZATA - OLSZEWSKI, ROBERT - PRZESNIAK-WELENC, MARTA - ŁAPIŃSKI, MARCIN - CIEŚLIK, BARTŁOMIEJ
0662	ADVANCED CHARACTERIZATION OF MACROSCOPIC ENSEMBLES OF CNTS VILATELA, JUAN JOSÉ - FERNÁNDEZ, JUAN CARLOS - RIDRUEJO, ÁLVARO
0666	FACTORS INFLUENCING THE ANTIBACTERIAL ACTIVITY OF GRAPHENE OXIDE BEYOND MATERIALS PHYSICOCHEMISTRY KARAHAN, H. ENIS - BAE, TAE-HYUN - XU, CHENJIE - WEI, JUN - CHEN, YUAN
0673	STUDY OF ION TRANSMISSION IN ELECTROLYTE OF GRAPHENE QUANTUM DOTS UNDER ULTRAVIOLET LIGHT WANG, XIAOMIN - CHEN, YINGPING - TIAN, ZHEN - ZHANG, HAIXIA - ZHANG, DING
0674	REMOVAL EFFICIENCY OF METHYL IODIDE BY KI/TEDA IMPREGNATED ACTIVATED CARBONS IN THE CONTEXT OF NUCLEAR INDUSTRY CHEBBI, MOUHEB - GERENTE, CLAIRE - LE COQ, LAURENCE - MOKILI, MARCEL - MONSANGLANT-LOUVET, CÉLINE
0675	THE INFLUENCE OF GIC STAGE NUMBER AND EXFOLIATION TEMPERATURE ON THE GAS PERMEABILITY OF GRAPHITE FOIL IVANOV, ANDREY - MALAKHO, ARTEM - MAKSIMOVA, NATALIA - IONOV, SERGEY - SHORNIKOVA, OLGA - AVDEEV, VIKTOR
0679	COMMERCIAL CARBON MOLECULAR SIEVES AS A HIGH PERFORMANCE ANODE FOR SODIUM-ION BATTERIES ZHANG, SIWEI
0685	RECENT ADVANCES IN SYNTHESIS AND CHARACTERIZATION OF CONTINUOUSLY SPUN CARBON NANOTUBE FIBER FROM THE GAS PHASE. RODILES, XAQUIN - VILA, MARIA - ALEMAN, BELEN - REGUERO, VICTOR - VILATELA, JUAN JOSE
0688	FLEXIBLE CARBON AEROGELS BY SPHERE TEMPLATING SALIHOVIC, MIRALEM - HÜSING, NICOLA - ELSAESSER, MICHAEL
0694	OVERCOMING THE MASS TRANSPORT LIMITATIONS OF IONIC LIQUIDS IN CO2 CHEMICAL ABSORPTION BY USING ENCAPSULATED IONIC LIQUIDS MOYA, CRISTIAN - SANTIAGO, RUBEN - LEMUS, JESUS - MORENO, DANIEL - ALONSO-MORALES, NOELIA - GILARRANZ, MIGUEL ANGEL - PALOMAR, JOSE
0697	CARBON FIBERS DERIVED FROM FRACTIONATED-SOLVATED LIGNIN OGALE, AMOD - JIN, JING - DING, JUNHUAN - THIES, MARK
0704	GRAPHENE AEROGELS MODIFIED WITH CO AND TI OXIDES AS ELECTROCHEMICAL CATALYSTS FOR OXYGEN EVOLUTION AND REDUCTION REACTIONS LUQUE CENTENO, JOSE MANUEL - MARTINEZ HUERTA, MARIA VICTORIA - CHEN, WEIYANG - SEBASTIAN, DAVID - YEUNG, KING LUN - LÁZARO ELORRI, MARIA JESUS
0705	DESIGN OF NANOSTRUCTURED CARBON MATERIALS VIA LASER ASSISTED SYNTHESIS TECHNIQUES GHIMBEU, CAMÉLIA - SOPORNYI, MIHAI - AXENTE, EMANUEL - SIMA, FELIX - NITA, CRISTINA
0706	DETERMINATION OF SURFACE ENERGY HETEROGENEITY OF CARBON-BASED NANOMATERIALS BY INVERSE GAS CHROMATOGRAPHY NADERI, MAJID - BURNETT, DAN
0711	COMPARISON BETWEEN IONIC LIQUIDS AND ENCAPSULATED IONIC LIQUIDS IN CO2 PHYSICAL ABSORPTION LEMUS, JESUS - SANTIAGO, RUBEN - MOYA, CRISTIAN - MORENO, DANIEL - ALONSO-MORALES, NOELIA - GILARRANZ, MIGUEL ANGEL - PALOMAR, JOSE

Oral Index

Code	Title
0713	DECORATION OF MEMBRANES VIA POLYDOPAMINE AND GRAPHENE OXIDE/AG NANOCOMPOSITES FOR CONDUCTIVE AND ANTIMICROBIAL PROPERTIES. LOPEZ-CAZARES, MARIA IRENE - PÉREZ-RODRÍGUEZ, FÁTIMA - CENTENO- SÁNCHEZ, MARISOL - CHÁZARO-RUIZ, LUIS FELIPE - RANGEL-MENDEZ, JOSE RENE
0714	EFFECT OF OCCURRENCE MODE OF AS ON VOLATILITY DURING PYROLYSIS PROCESS ZHOU, LINGMEI - CHU, MO
0717	CNT FIBERS AS MULTIFUNCTIONAL ELECTRODE MATERIAL FOR FLEXIBLE SUPERCAPACITORS. ELECTROCHEMICAL PROPERTIES AND APPLICATION. SENOKOS, EVGENY - RANA, MOUMITA - REGUERO, VICTOR - PENDASHTEH, AFSHIN - IGLESIAS, DANIEL - MARCHESAN, SILVIA - MARCILLA, REBECA - VILATELA, JUAN JOSE
0719	ECO-FRIENDLY AND FACILE SYNTHESIS OF CARBON/SIO2 HYBRID MATERIALS FOR LI-ION BATTERIES APPLICATIONS NITA, CRISTINA - FULLENWARTH, JULIEN - PARMENTIER, JULIEN - MONCONDUIT, LAURE - MATEI GHIMBEU, CAMELIA
0723	SYNTHESIS OF MULTIWALLED CARBON NANOTUBE-SUPPORTED CUCO2S4 FOR ENHANCED FENTON-LIKE CATALYSIS TAO, YOUSHENG - ZHANG, MAI - ANNAMALAI, KUNDRAKUDI P. - CHEN, TIANLU
0724	FINGER-LIKE CARBON-BASED NANOSTRUCTURES PRODUCED BY COMBUSTION OF SPAGHETTI OSSLER, FREDERIK - HETHERINGTON, CRISPIN
0729	NITRATE ION ADSORPTION ON ACTIVATED CARBONS WITH DIFFERENT TEXTURAL AND PHYSICOCHEMICAL CHARACTERISTICS HENRIETTE, MOUSSOUNDA BOUNDZANGA - BENOÎT, CAGNON - CHRISTINE, VAUTRIN-UL - SYLVIE, BONNAMY
0730	NEUTRON IMAGING AND SCATTERING BASED CHARACTERIZATION OF THE INTERACTION BETWEEN HYDROGENOUS SUBSTANCES AND SOOT MATERIALS OSSLER, FREDERIK - SANTODONATO, LOUIS - DHIMAN, INDU - FINNEY, CHARLES - ARMITAGE, DOUGLAS - BILHEUX, JEAN-CHRISTOPHE - DAEMEN, LUKE - RAMIREZ-CUESTA, ANIBAL - BILHEUX, HASSINA
0731	THE USE OF WOOD RESIDUES FROM THE BOREAL FOREST FOR THE PRODUCTION OF POROUS MATERIALS FOR ENVIRONMENTAL APPLICATIONS BRAGHIROLI, FLAVIA - BOUAFIF, HASSINE - NECULITA, CARMEN MIHAELA - KOUBAA, AHMED
0732	THE EFFECT OF OXIDATION TREATMENT ON THE ADSORPTION OF AMMONIA ON CARBON ELECTRODES FOR ELECTROCHEMICAL CAPACITORS SLESINSKI, ADAM - FIC, KRZYSZTOF - FRACKOWIAK, ELZBIETA
0733	RECOVERING PORES SIZE DENSITY DISTRIBUTION OF ACTIVATED CARBON BY REGULARIZING AND HRTEM IMAGES ANALYSIS PROCEDURES ALVAREZ, AMAURY CRUZ - JEANNE-ROSE, VALÉRIE - MONTRESOR PASSE-COUTRIN, NADY - GASPARD, SARRA
0735	THERMAL EXFOLIATION OF GRAPHITE OXIDE FILMS – THERMAL HAZARD OR AN OPPORTUNITY FOR LARGE-SURFACE-AREA GRAPHENE-BASED MATERIALS? DECASTILHO, CINTIA - BERHANE, SARA - HURT, ROBERT - KULAOTS, INDREK
0739	KINETIC STUDY OF METHANOL DEHYDRATION OVER ZRO2 SUPPORTED-ACTIVATED CARBONS PALOMO, JOSE - RODRÍGUEZ-CANO, MIGUEL ÁNGEL - ROSAS, JUANA MARÍA - RODRÍGUEZ-MIRASOL, JOSÉ - CORDERO, TOMÁS
0741	CARBON MOLECULAR SIEVES FROM FRUIT WASTE FOR BIOGAS SEPARATION AMAYA, ALEJANDRO - TANCREDI, NESTOR
0743	HOW THE STRUCTURE OF THE CARBON SUPPORT INFLUENCE ON STRUCTURE AND DURABILITY OF DEPOSITED PLATINUM NANOPARTICLES? PAWLYTA, MIROSLAWA - LISZKA, BARBARA - JURKIEWICZ, KAROLINA - SZADE, JACEK - BLACHA-GRZECHNIK, ALICJA
0744	WATER SORPTION PHENOMENON ON SOLID PARTICLES EMITTED DURING A FIRE: DETERMINATION OF THE INFLUENCING PHYSICO-CHEMICAL PARAMET LAURA, LINTIS - FRANÇOIS-XAVIER, OUF - ALEXIS, COPPALLE - CÉCILE, VALLIÈRES
0746	GEOCHEMICAL AND PHYSICOCHEMICAL CHARACTERISTICS OF SOME COLOMBIAN COALS AND THEIR RELATIONSHIP WITH THEIR COAL METHANE CONTEN CASTRO, ADRIANA - PARRA, SERGIO D. - GUERRERO, JAVIER - RINCÓN, MARCO A. - ROMERO, MANUEL H. - AGÁMEZ-PERTUZ, YAZMIN Y. - DIAZ, JOSE DE J. - CADENA, ARIEL O.
0756	LASER REDUCTION OF OXIDE GRAPHENE UNDER AMBIENT CONDITION EVLASHIN, STANISLAV - SAFONOV, ALEXANDER - FEDOROV, FEDOR - SVYAKHOVSKIY, SERGEY - DAGESYAN, SARKIS - DYAKONOV, PAVEL - MANKELEVICH, YURI - SUETIN, NIKOLAY - AKHATOV, ISKANDER
0757	HIGH CAPACITY CARBON NANOTUBE FIBER FOR BINDER-FREE LI-ION BATTERY ANODES BEYOND-GRAPHITE CINTORA JUAREZ, DANIEL - AVVARU, VENKATA SAI - ALEMÁN, BELÉN - VILATELA, JUAN JOSE - ETACHERI, VINODKUMAR
0758	BANDED IRON FORMATION FOR EFFICIENT NITROGEN-DOPED CARBON NANOTUBES PRODUCTION. JIMENEZ-RAMIREZ, LUIS E. - FAJARDO-DIAZ, JUAN L. - VERMA, SANJEET J. - LOPEZ URIAS, FLORENTINO - MUÑOZ SANDOVAL, EMILIO

Code	Title
0765	SOME PROPERTIES OF WOOD PLASTIC NANOCOMPOSITES REINFORCED WITH MULTI WALLED CARBON NANOTUBES KAYMAKCI, ALPEREN - AYRILMIS, NADIR - OZKAN, OSMAN EMRE - GULEC, TURKER
0777	PREPARATION OF ISOTROPIC PITCH-BASED CARBON FIBER FROM PYROLYTIC OILS OF BIOMASS AND POLYSTYRENE WITH ETHYLENE BOTTOM OIL ÖZSIN, GAMZENUR - PÜTÜN, AYŞE EREN - NAKABAYASHI, KOJI - MIYAWAKI, JIN - YOON, SEONG-HO
0780	APPLE TREE BRANCHES ACTIVATED CARBONS FOR THE REMOVAL OF ATENOLOL MARQUES, SUSANA C. R. - MESTRE, ANA S. - MACHUQUEIRO, MIGUEL - GOTVAJN, ANDREJA - MARINŠEK, MARJAN
0784	IN SITU NMR STUDY OF IONS TRANSFER DURING THE CHARGE/DISCHARGE OF CARBON-BASED SUPERCAPACITORS OUKALI, GHENIMA - DESCHAMPS, MICHAEL - PINERO, ENCARNACION RAYMUNDO - SALAGER, ELODIE
0794	MECHANISMS OF HYDROGEN STORAGE ON PALLADIUM DOPED NANOPOROUS CARBONS LOPEZ, MARIA J. - JUARISTI, J.IÑAKI - ALDUCIN, MAITE - GRANJA-DELRIO, ALEJANDRA - ALONSO, JULIO A.
0799	NOVEL APPROACHES FOR HYDROCHARS SULFONATION DUARTE, TIAGO - MARTINS, LUÍSA - CARVALHO, ANA
0800	TWO SPRAYER CVD SYNTHESIS OF NITROGEN-DOPED CARBON NANOTUBE SPONGE MUÑOZ-SANDOVAL, EMILIO - FAJARDO-DÍAZ, JUAN LUIS - SÁNCHEZ-SALAS, ROQUE - CORTÉS-LÓPEZ, ALEJANDRO JAVIER - LÓPEZ-URÍAS, FLORENTINO
0802	FUNCTIONALIZATION WITH A PROBE MOLECULE TOWARDS THE BASAL PLANE OF THE GRAPHENE OXIDE PIÑEIRO GARCIA, ALEXIS - VEGA DIAZ, SOFIA MAGDALENA - TRISTÁN LÓPEZ, FERDINANDO - FIERRO GONZÁLEZ, JUAN CARLOS - SEMETEY, VINCENT - MENESES RODRÍGUEZ, DAVID
0808	COATING CONVENTIONAL GLASS AND POLYCARBONATE WITH TRANSPARENT GRAPHENE-TYPE LAYERS VERNET-GARCÍA, JOAN - DOMINGO-GARCÍA, MARÍA - PÉREZ-MENDOZA, MANUEL - GARCÍA-GALLARÍN, CELESTE - MELGUIZO, MANUEL
0810	RICE HUSK-DERIVED C/C COMPOSITE HONEYCOMB MONOLITHS FOR ELIMINATION OF LIPOPOLYSACCHARIDE ENDOTOXIN MIKHALOVSKY, SERGEY - JANDOSOV, JAKPAR - MISRA, ANIL - ALAVIJEH, MO - HOWELL, CAROL - MANSUROV, ZULKHAIR - SILVESTRE-ALBERO, JOAQUIN - MIKHALOVSKA, LYUBA - TENNISON, STEVE - MIKHALOVSKY, SERGEY
0811	OXYGEN REDUCTION REACTION IN NITROGEN-DOPED MULTIWALL CARBON NANOTUBES SÁNCHEZ-SALAS, ROQUE - LÓPEZ URÍAS, FLORENTINO - MIRANDA HERNÁNDEZ, MARGARITA - CUENTAS GALLEGOS, ANA KARINA - MUÑOZ SANDOVAL, EMILIO
0819	FLUORINATION OF NUCLEAR GRAPHITE IG-110 IN MOLTEN FLIBE SALT AT 700 °C SCARLAT, RALUCA - WU, HUALI
0827	OUTSTANDING ELECTROCHEMICAL PERFORMANCES OF HIGHLY N- AND O-DOPED CARBONS DERIVED FROM PINE TANNIN SANCHEZ-SANCHEZ, ANGELA - IZQUIERDO, MARÍA TERESA - MATHIEU, SANDRINE - GONZALEZ-ALVAREZ, JULIA - CELZARD, ALAIN - FIERRO, VANESSA
0828	HIGH SURFACE-AREA, TANNIN-DERIVED, ORDERED MESOPOROUS CARBONS AS ELECTRODES OF SUPERCAPACITORS CASTRO-GUTIÉRREZ, JIMENA - SANCHEZ-SANCHEZ, ANGELA - IZQUIERDO, MARÍA TERESA - CELZARD, ALAIN - FIERRO, VANESSA
0829	SURFACTANT-WATER-ASSISTED MECHANOCHEMICAL MESOSTRUCTURACTION OF TANNINS TO SYNTHESISE OMCS FIERRO, VANESSA - CASTRO-GUTIERREZ, JIMENA - SANCHEZ-SANCHEZ, ANGELA - CELZARD, ALAIN
0837	CARBON NANOCONES AS MULTI-MODE PROBES FOR NEARFIELD MICROSCOPIES PAREDES, GERMERCY - SEINE, GREGORY - PALLEAU, ETIENNE - DELMAS, MATHIEU - RESSIER, LAURENCE - COURS, ROBIN - MASSEBOEUF, AURELIEN - ONDARCUHU, THIERRY - PIAZZA, FABRICE - MONTHIOUX, MARC
0841	INTERACTIONS BETWEEN GRAPHITE FLUORIDES AND FEMTOSECOND LASER : FROM CONTROLLED REDUCTION TO EFFICIENT EXFOLIATION HERRAIZ, MICHAEL - BATISSE, NICOLAS - DUBOIS, MARC - PEYROUX, JEREMY
0842	NEW COMPOSITE MATERIALS BASED ON POLYETHYLENE AND MULTI-WALLED CARBON NANOTUBES MODIFIED BY CO NANOPARTICLES KAZAKOVA, MARIYA A. - SEMIKOLENOVA, NINA V. - KOROVIN, EVGENIY YU. - KUZNETSOV, VLADIMIR L. - SUSLYAEV, VALENTIN I. - MATSKO, MIKHAIL M. - ZAKHAROV, VLADIMIR A.
0845	3D CARBON MICROSTRUCTURES BY CARBONIZATION OF FEMTOSECOND LASER POLYMERIZED PHOTORESISTS BATISSE, NICOLAS - HERRAIZ, MICHAEL - NASR, MARYLINE - DUBOIS, MARC
0850	TUNABLE NANOPARTICLES IN ALUMINUM CARBIDE-DERIVED CARBON FOR ENHANCED ACID GAS ADSORPTION MORAN, COLTON - MARTI, ROBERT - HAYES, SOPHIA - WALTON, KRISTA
0856	STARENE®: STARCH/GRAPHITE MESOPOROUS CARBON MONOLITH EDLC ELECTRODES SHUTTLEWORTH, PETER S. - MUÑOZ, ANDREA - BUDARIN, VITALY - CLARK, JAMES - LAZAROV, VLADO - LARI, LEONARDO - SALAVAGIONE, HORACIO - ELLIS, GARY - MORALES, ENRIQUE

Oral Index

Code	Title
0860	SPECTROSCOPIC SIGNATURES OF AQUEOUS DISPERSIONS OF SINGLE LAYER GRAPHENE AND SINGLE-WALLED CARBON NANOTUBES GEORGE, BEPETE - CARLOS, DRUMMOND - ALAIN', PENICAUD - ERIC, ANGLARET
0864	IMPROVEMENT OF BIODEGRADABILITY USING BIOCHAR IN DRY BIOTOILET YOSHIZAWA, SHUJI - TANAKA, SATOKO - HASHIKAWA, YUKO - IWAO, SHINICHI
0870	EMERGING CARBON ARCHITECTURES FOR SAFER ENERGY STORAGE DEVICES POL, VILAS
0873	ELECTRON CONDUCTING CARBON-BASED CEMENT ELECTRODES: EXPANDING CONCRETE STRUCTURAL ELEMENTS FUNCTIONALITIES IOANNIDOU, KATERINA - CHANUT, NICOLAS - DIVOUX, THIBAUT - BACKOV, RÉNAL - ULM, FRANZ-JOSEF - PELLENQ, ROLAND
0874	HEMATIN SUPPORTED ON GRAFENIC MATERIALS: A BIOMIMETIC CATALYSTS FOR THE POLYANILINE SYNTHESIS ROMERO-GRACIA, JORGE - RAMÍREZ-HERNÁNDEZ, ABRIL-VIANNET - MARTÍNEZ-CARTAGENA, MANUEL - FERNÁNDEZ-TAVIZÓN, SALVADOR - RODRÍGUEZ, JOELIS - LEDEZMA-PÉREZ, ANTONIO - DÍAZ-BARRIGA, ENRIQUE
0876	INTERLAYER STUDY OF GRAPHENE STACKS ZHAO, PEI - WANG, YUNLU - ZHANG, XUEWEI - WANG, HONGTAO
0879	THREE-DIMENSIONAL BILAYER GRAPHENE STRUCTURES PRODUCED BY PASSAGE OF AN ELECTRIC CURRENT HARRIS, PETER
0881	RAPID GROWTH OF METAL OXIDE-CARBON HYBRID NANOWIRES FOR SUPERCAPACITORS LIU, YANG - ZHANG, XINYU - XIN, JOHN - ZHANG, CHAO
0882	NEW ADVANCES IN THE DEVELOPMENT OF CARBONACEOUS SORBENTS TO RETAIN MERCURY IN THE GAS PHASE ANTUÑA-NIETO, CRISTINA - RODRIGUEZ, ELENA - LOPEZ-ANTON, MARIA ANTONIA - GARCIA, ROBERTO - MARTINEZ-TARAZONA, MARIA ROSA
0889	HIGH-TEMPERATURE TREATMENT OF CHAR OBTAINED FROM KRAFT LIGNIN BY PYROLYSIS ALONSO-MORALES, NOELIA - HERAS, FRANCISCO - ESER, SEMIH - TOBAJAS, MONTSERRAT - RODRIGUEZ, JUAN - GILARRANZ, MIGUEL ANGEL
0891	COMPARISON OF RAMAN SPECTRA OF PARTICULATE MATTER PRODUCED BY RENEWABLE DIESEL AND FOSSIL DIESEL ON DIFFERENT COMBUSTION SYST CADRAZCO, MARLON - AGUDELO, JOHN - SANTAMARÍA, ALEXANDER
0892	PHYSICOMECHANICAL PROPERTIES AND IN VITRO BIOLOGICAL PERFORMANCE OF KERATIN/BACTERIAL CELLULOSE NANOFIBROUS COMPOSITES FOR WOUND DRESSING AZARNIA, AMIR - TAMJID, ELNAZ - ESLAHI, NILOOFAR - SIMCHI, ABDOLREZA
0893	PREPARATION AND CHARACTERIZATIONS OF CU/CARBON SPHERES MATERIAL. APPLICATION FOR CO2 PHOTO-ELECTROCHEMICAL REDUCTION AMEL, BOUDJEMAA - SOUAD, BENAMMAR - NEZZAL, GANIA - SANTIAGO, GÓMEZ-RUIZ - DALILA, MEZIANE - KHALDOUN, BACHARI - NEIL, COVILLE
0904	ENHANCED ELECTROCHEMICAL PERFORMANCE OF DOPED REDUCED GRAPHENE OXIDE BASED BIMETALLIC NI/FE NANOHYBRIDS FOR RAPID DECHLORINAT SAHU, RAMA SHANKER - DOONG, RUEY-AN - LIN, FANG-HSIN
0905	ACTIVATED CARBON/FERRITE NANOPARTICLES COMPOSITE AS A MAGNETIC DRUG DELIVERY SYSTEM RIOS HURTADO, JORGE CARLOS - MUZQUIZ RAMOS, ELIA MARTHA - MONTAÑO FIGUEROA, ANA GABRIELA - FUENTES, ERIKA - RANGEL MENDEZ, JOSE RENE - ESCALANTE IBARRA, GRISELDA
0914	IMPROVEMENT OF CO2 CAPTURE BY FE DOPED CARBONIZED WOOD HATA, TOSHIMITSU - HONMA, SENSHO - ONISHI, YOSHIKAZU - IDE, ISAMU - BONNAMY, SYLVIE - BRONSVELD, PAUL - TANAKA, HIDEKI
0920	TRACE CALCIUM INDUCED CATALYTIC ACTIVATION ACHIEVING PORE REORGANIZATION OF COAL CONVERTED POROUS CARBONS FOR EFFICIENT CO2 A WANG, LIJIE - SUN, FEI - GAO, JIHUI - LI, LONGXIN - QIN, YUKUN
0921	NOVEL CARBON FOAMS FROM CURED PHENOLIC RESINS PREPARED BY REPRECIPITATION AND EMULSIFICATION OF RESIN PRECURSOR DIAZ AUNON, JOSE ANTONIO - KOZYNCHENKO, OLEKSANDR
0923	ALKANE DEHYDROGENATION ON NANOCARBON: FROM REACTION MECHANISM TO CATALYST DESIGN QI, WEI
0924	BIOMASS-DERIVED ELECTRODES FOR BIOELECTROCHEMICAL SYSTEMS MARZORATI, STEFANIA - BAHDANCHYK, MAKSIM - GOGLIO, ANDREA - TRASATTI, STEFANO - CRISTIANI, PIERANGELA - SCHIEVANO, ANDREA
0927	INDUCING COKE FORMING ABILITY IN VICTORIAN BROWN COAL MOLLAH, M MAMUN - MARSHALL, MARC - JACKSON, W ROY - CHAFFEE, ALAN L

Code	Title
0930	FUCNTIONALIZATION OF CARBON NANOSTRUCTURES FOR THE PRODUCTION OF H_2O_2 AND SUPERCAPACITORS IGLESIAS, DANIEL - GIULIANI, ANGELA - SENOKOS, EVGENY - MELCHIONNA, MICHELE - VILATELA, JUAN JOSÉ - FORNASIERO, PAOLO - PRATO, MAURIZIO - MARCHESAN, SILVIA
0932	N-DOPED CARBON GELS BASED ON MELAMINE-LIGNIN-RESORCINOL MIXTURES FOR SUPERCAPACITOR APPLICATION ZAPATA BENABITHE, ZULAMITA - DIOSSA, GIOVANNA - YEPES LÓPEZ, JESSICA - CÓRDOBA MONTES, JOHN ANDRÉS - QUINTANA, GERMÁN
0934	FILAMENT PRINTING OF GRAPHENE-BASED INKS INTO SELF-SUPPORTED 3D ARCHITECTURES MOYANO, JUAN JOSE - BELMONTE, MANUEL - MIRANZO, PILAR - OSENDI, MARIA ISABEL
0935	DEVELOPMENT OF ACTIVATED PHENOL-FORMALDEHYDE BEADS WITH BAMBOO TAR ADDITION FOR ORGANIC VAPOR ADSORPTION CHEN, YU-TING - HUANG, YING-PIN - ROOD, MARK J. - HSI, HSING-CHENG
0942	ELECTROCONDUCTIVE CARBON BIOFILTERS FOR EFFICIENT AND SUSTAINABLE TREATMENT OF URBAN WASTEWATER PRADO, AMANDA - BERENGUER, RAUL - ESTEVE-NÚÑEZ, ABRAHAM
0946	VALUE ADDITION TO LOCAL BIOMASS: SYNTHESIS OF ACTIVATED CARBON FROM BABASSU MESOCARP AND ITS ENVIRONMENTAL APPLICATIONS GHOSH, ANUPAM - MARIA DA SILVA SANTOS, ARIANE - DO AMARAL RAZZINO, CLAUDIA - DELFINO DE ARAGÃO REIS, ALEXANDRE - COSTA SILVA, RUBENS - FUJISAWA, KAZUNORI - PASTOR FERREIRA, ODAIR - O LOBO, ANDERSON - TERRONES, MAURICIO - CRUZ VIANA, BARTOLOMEU
0947	SINGLE WALLED CARBON NANOTUBE FIELD EFFECT TRANSISTOR AS MEMORY DEVICE BARGAOUI, YASMINE - TROUDI, MANEL - BONDAVALLI, PAOLO - SGHAIER, NABIL
0950	GRAPHENE/NANOROD COMPOSITES FOR ENZYMATIC AND NON-ENZYMATIC ELECTROCHEMICAL BIOSENSING: APPLICATION TO GLUCOSE DETECTION MAZAHERI, MOZHDEH - AASHURI, HOSSEIN - SIMCHI, ABDOLREZA
0953	CARBOHYDRATE COATINGS FOR BIOFOULING MITIGATION: DIAZONIUM BASED SACCHARIDE COATINGS FOR IMPROVED ELECTROCHEMICAL SENSORS MYLES, ADAM - BEHAN, JAMES - SCANLAN, EOIN - COLAVITA, PAULA
0955	EFFECT OF GRAPHENE OXIDE NANOSHEETS ON THE PHYSICOCHEMICAL AND BIOLOGICAL PERFORMANCE OF ELECTROSPUN POLYMERIC NANOFIBERS MAHMOUDI, NAFISHEH
0958	COMPUTATIONAL STUDY OF A MENSHUTKIN CHEMICAL REACTION INSIDE DOPED CARBON NANOTUBES MAGALHAES, ALEXANDRE - ALVES, INES
0964	EFFECT OF HYDROTHERMAL PRETREATMENT ON THE POROSITY OF PYROLYTIC CHAR VURAL, E. GÖZDE - KILIÇ, MURAT - ÖZSİN, GAMZENUR - APAYDIN VAROL, ESIN
0965	PRODUCTION OF ACTIVATED CARBON FROM ASPARAGUS OFFICINALIS L FOR 2,4-DICHLOROPHENOL ADSORPTION PRESENT IN AQUEOUS SOLUTION ANDRADE-ESPINOSA, GUILLERMO - BARCENAS-TORRES, JOSÉ DIEGO - SÁNCHEZ-ORTIZ, OMAR AUGUSTO - BERBER-MENDOZA, MARIA SELENE - ALVARADO-FLORES, JOSÉ JUAN
0969	CARBON IN THE SOIL: PREPARATION AND CHARACTERIZATION OF BIOCHAR FROM BEECH WOOD SAWDUST AYGÜN, BÜŞRA - APAYDIN-VAROL, ESIN
0970	DOPED GRAPHENE BASED ELECTROCATALYSTS FOR HYDROGEN EVOLUTION REACTION LEMES, GIOVANNI - SEBASTIÁN, DAVID - LÁZARO, MARÍA JESÚS - PASTOR, ELENA
0972	ELECTROCHEMICAL PERFORMANCE IN AQUEOUS AND ORGANIC MEDIUM OF ACTIVATED CARBONS FROM AFRICAN PALM KERNEL SHELLS ARANGO, DANIEL I. - ZAPATA-BENABITHE, ZULAMITA - ARENAS, ERIKA C.
1034	PREPARATION AND CHARACTERISATION OF PAN-BASED CARBON NANOFIBRES WITH GRAPHENE NANOPLATELETS BY USING PRESSURISED GYRATION WU, XIAOWEN - MAHALINGAM, SUNTHARAVATHANAN - AMIR, AMALINA - PORWAL, HARSHIT - REECE, MIKE - COLOMBO, PAOLO - EDRISINGHE, MOHAN

Posters Index

Posters Index

Code	Title
0007	CA-P COATING PREPARED BY SUPERSONIC ATMOSPHERIC PLASMA SPRAYING COMBINING MICROWAVE-HYDROTHERMAL METHODS SU, YANGYANG - LI, KEZHI - ZHANG, LEILEI
0010	THERMAL AND ELECTRICAL PROPERTIES BASED ON REINFORCED CARBON/POLYAMIDE-6 COMPOSITES PARK, SOOYOUL - KIM, SONGHUI - PARK, DONGJUN - LEE, BYUNGMIN
0011	CATALYTIC OXIDATION OF LEAN METHANE SIDANA, SMARTHVEER
0020	ASYMMETRIC VANADIUM DIOXIDE//CARBONIZED (PANI) GRAFTED WITH IRON NANOPARTICLES (C-FP) SUPERCAPACITOR FOR HIGH ENERGY DENSITY NDIAYE, NDEYE MATY - MADITO, MOSHAWE - NGOM, BALLA DIOP - DANGBEGNON, JULIEN - MASIKHWA, TSHIFHIWA - MANYALA, NCHOLU
0022	SYNTHESIS AND INVESTIGATION OF TWO-DIMENSIONAL TI3C2-MN3O4 NANOCOMPOSITE FOR ELECTROCHEMICAL CAPACITOR ELECTRODES OYEDOTUN, KABIR - MOMODU, DAMILOLA - MIRGHNI, ABDULMAJID - MASIKHWA, TSHIFHIWA - KHALEED, ABUBAKAR - MANYALA, NCHOLU
0024	HYBRID ASYMMETRIC CAPACITOR BASED ON CO3(PO4)2.4H2O/GRAPHENE FOAM COMPOSITE WITH ENHANCED CAPACITY AND LONG CYCLE LIFE MIRGHNI, ABDULMAJID A. - MOMODU, DAMILOLA - OYEDOTUN, KABIR O. - DANGBEGNON, JULIEN K. - MANYALA, NCHOLU
0028	INFLUENCE OF DIFFERENT EXPERIMENTAL PARAMETERS ON THE CHARACTERIZATION OF ACTIVATED CARBONS BY MEANS OF GAS ADSORPTION FERNÁNDEZ VELASCO, LETICIA - DEVOS, ANNELEEN - LODEWYCKX, PETER
0036	LARGE-AREA GRAPHENE MEMBRANES AND MESOPOROUS DEIONIZATION ELECTRODES FOR WATER DESALINATION GUPTA, SANJU - HENSON, ALEX - EVANS, BRENDAN
0038	GRAPHENE-CARBON NANOTUBE AEROGEL AS ORGANIC THERMOELECTROCHEMICAL ENERGY HARVESTERS GUPTA, SANJU - MEEK, ROMNEY - EVANS, BRENDAN
0041	HIGHLY ORDERED POROUS CARBONS BY MICROWAVE SYNTHESIS CANAL-RODRÍGUEZ, MARIA - VILLANUEVA, SARA F. - MONTES-MORAN, MIGUEL - ARENILLAS, ANA - MENENDEZ, J. ANGEL
0042	CO2 CAPTURE BY ORDERED MESOPOROUS/MICROPOROUS CMK-3 CARBON: PORE SIZE EFFECT ON KINETICS OF CO2 ADSORPTION VOROKHTA, MARYNA - MORAVKOVA, JAROSLAVA - PILAR, RADIM
0043	INFLUENCE OF THE SURFACE CHARACTERISTICS OF CARBON STRUCTURED MONOLITHS ON NANOPARTICLE GOLD DEPOSITION IZQUIERDO, MARIA TERESA - GÓMEZ-GIMÉNEZ, CARMEN - BALLESTERO, DIEGO - RUBIO, BEGOÑA
0049	HIERARCHICAL CO3O4/HOLLOW CARBON NANOSPHERE@CARBON NANOFIBERS HYBRIDIZED CATHODE MATERIALS FOR LITHIUM-SULFUR BATTERY VIA ELE HE, LEI - ZHOU, YING - QIU, JIESHAN
0051	DEACTIVATION PATHWAYS OF SULFONATED CARBON CATALYSTS IN HYDROTHERMAL REACTION ENVIRONMENTS SCHOLZ, DAVID - KRÖCHER, OLIVER - VOGEL, FRÉDÉRIC
0063	SYNTHESIS OF MAGNETITE NANOPARTICLES OBTAINED BY SOLUTION COMBUSTION METHOD LESBAYEV, AIDOS - SMAGULOVA, GAUKHAR - NIKOLAY, PRIKHODKO - MANSUROV, ZULKHAIR - USTAYEVA, GAUKHAR
0065	KINETICS OF WATER ADSORPTION ON ACTIVATED CARBONS BOUTILLARA, YASMINE - LODEWYCKX, PETER - RICHELET, LUCAS - LÉONARD, ANGÉLIQUE
0066	ULTRASONIC-ASSISTED REDUCTION TO PREPARE SUPPORTED PD/C CATALYST FOR SUZUKI-MIYAURA COUPLING REACTION WU, WEI
0067	HIGHLY-DISPERSED PD NANOPARTICLES SUPPORTED ON MULTI-WALLED CARBON NANOTUBES FOR SUZUKI COUPLING REACTION LIU, YANG - BAI, XUEFENG
0069	ROLE OF ACTIVATING AGENTS AND TEMPERATURE IN THE DIELECTRIC PROPERTIES OF BIOMASS FOR PREPARATION OF ACTIVATED CARBON BY MICR DURAN, GABRIELA - BINNER, ELEANOR - HERNANDEZ, VIRGINIA
0070	CHARACTERIZATION OF POLYMER CARBON SIEVES BETZ, WILLIAM - PENA DUQUE, LEIDY - KEELER, MICHAEL - JONES, JAY - FRANTZ, CURTIS - SMITH, BRITTANY
0081	CONTRIBUTION OF NANODIAMONDS AND NANOONIONS TO CHARGE STORAGE IN SUPERCAPACITORS LEDERER, INGO - MUZHA, ANDREAS - KRUEGER, ANKE - REICHENAUER, GUDRUN

Code	Title
0085	HIGHLY EFFICIENT REMOVAL OF PHARMACEUTICALS FROM WATER BY ADSORPTION ONTO WELL-DEFINED CARBIDE-DERIVED CARBONS ALVAREZ-TORRELLAS, SILVIA - MUNOZ, MACARENA - GLÄSEL, JAN - MARTINEZ DE PEDRO, ZAHARA - ETZOLD, BASTIAN J.M. - CASAS, JOSE ANTONIO
0092	INFLUENCE OF GLYME-BASED ELECTROLYTES COMPOSITION ON THE ANODIC PERFORMANCE OF GRAPHITIC CARBON FOAMS IN SODIUM-ION BATTERIES CAMÉAN, IGNACIO - RODRIGUEZ-GARCÍA, JORGE - GARCIA SUAREZ, ANA B.
0094	HIERARCHICAL POROUS CARBON FOR ULTRACAPACITORS ROH, KWANG CHUL
0097	LARGE-SCALE, LOW-COST, CONTROLLABLE AND TEMPLATE-FREE PREPARATION OF HOLLOW SPHERICAL ACTIVATED CARBON WITH A HONEYCOMB POROU GUAN, TAOTAO - ZHANG, GUOLI - LI, YONGXIN - ZHAO, JIANGHONG - WANG, JIANLONG - LI, KAIXI
0098	GRAPHITIZATION OF ACTIVATED CARBONS VIA VACUUM ANNEALING CHRISTIANS, HAUKE - PRÖSSL, CAROLIN - GLÄSEL, JAN - ETZOLD, BASTIAN J.M.
0099	CARBIDE-DERIVED CARBONS AS STATIONARY PHASE IN HIGH-PERFORMANCE LIQUID CHROMATOGRAPHY GEISSLER, MIRIAM - CHRISTIANS, HAUKE - GLÄSEL, JAN - KASPEREIT, MALTE - ETZOLD, BASTIAN J. M.
0102	SYNTHESIS OF METHANE BY HYDROGENATION OF CO2 USING CATALYSTS BASED NI DOPED WITH MG AND/OR CE. MONZON, ANTONIO - CAZAÑA, FERNANDO - SEBASTIÁN, VICTOR - LATORRE, NIEVES - ROMEO, EVA - ROYO, CARLOS - TARIFA, PILAR
0106	WORK FUNCTION OF GRAPHENE SURFACES AS A DESCRIPTOR FOR EVALUATION OF BACTERIA ADHESION DUCH, JOANNA - GOLDA-CEPA, MONIKA - KOTARBA, ANDRZEJ
0108	SHORT CARBON FIBERS AS ADDITIVE FOR SUPERCAPACITOR ELECTRODES KONRAD, KROIS - JAN, GLÄSEL - BASTIAN, ETZOLD
0119	SYNTHESIS OF GRAPHITIC CARBONS BY CATALYTIC GRAPHITIZATION FOR THE APPLICATION IN LITHIUM ION BATTERIES VANESSA, HANHART - MARTIN, WINTER - TOBIAS, PLACKE
0124	SYNTHESIS OF HIERARCHICAL TIO2/MWCNT COMPOSITES FOR VOCS ABATEMENT FERNÁNDEZ CATALÁ, JAVIER - BERENGUER MURCIA, ÁNGEL - CAZORLA AMORÓS, DIEGO CAZORLA-AMORÓS
0126	SYNTHESIS OF POROUS DOPED CARBON MATERIALS FROM POLYSACCHARIDES AMANTIA, DAVID - MARTÍNEZ-CRESPIERA, SANDRA - PEREZ TEJADA, RAQUEL - MACQUARRIE, DUNCAN - BUDARIN, VITALIY - ÇAVUS, GOKHAN - AUCHER, CHRISTOPHE
0127	NOX ADSORPTION PERFORMANCE OF DIFFERENT CHARCOALS PHYSICALLY ACTIVATED WITH WATER VAPOR SAGER, UTA - GHOUMA, IMEN - JEGUIRIM, MEJDI - DÄUBER, ECKHARD - LIMOUSY, LIONEL - BENNICI, SIMONA - GHIMBEU, CAMÉLIA MATEI - LIGOTSKI, ROMAN - SCHMIDT, FRANK - OUEDERNI, ABDELMOTTALEB
0136	FACILE SYNTHESIS OF CHIRAL CARBON DOTS DERIVED FROM TRYPTOPHAN WEI, YINGYING - CHEN, LIN - WANG, JUNLI - YU, SHIPING - YANG, YONGZHEN - LIU, XUGUANG - HU, LANQING
0153	INFLUENCE OF P/FE COMPLEXES FORMED ON NANOPOROUS CARBONS UPON THE PHOTOCATALYTIC DEGRADATION OF METHYLENE BLUE MATOS, JUAN - MONTAÑA, RICMARY - ARCIBAR-OROZCO, JAVIER - RANGEL-MÉNDEZ, JOSE R.
0159	REMOVAL OF DICLOFENAC, IBUPROFEN AND PARACETAMOL BY NITROGEN-DOPED CARBON SPONGES ORTIZ LÓPEZ, TANIA - LÓPEZ URÍAS, FLORENTINO - MUÑOZ SANDOVAL, EMILIO
0160	BORON-NITROGEN HETERODOPING IN MULTIWALLED CARBON NANOTUBES OLVERA MARTÍNEZ, JOSÉ AGUSTÍN - LÓPEZ URÍAS, FLORENTINO - MUÑOZ SANDOVAL, EMILIO
0167	OPTO-THERMAL TECHNIQUE FOR MEASURING THERMAL CONDUCTIVITY OF POLYACRYLONITRILE BASED CARBON FIBERS JANG, DAWON - LEE, DONGSU - JOH, HAN-IK - LEE, SUNGHO
0169	A COMPARATIVE STUDY ON THE LATERAL SIZE EFFECT OF THE MULTI-LAYERED GRAPHENE ON THE ELECTRICAL CONDUCTIVITY AN, JUNG-CHUL - LEE, EUN JUNG - YOON, SO-YOUNG - KIM, YONG-JUNG - LEE, SEONG-YOUNG
0170	A PDMS BASED 3 OMEGA SENSOR FOR THERMAL CHARACTERIZATION OF POLYMER COMPOSITES WITH CARBON FIBER OH, DONG-WOOK
0178	PREPARATION OF SILICON@ POROUS CARBON COMPOSITES BY DIFFERENT CARBON SOURCES FOR LITHIUM BATTERY ANODES FU, DONGJU - MIN, JIE - ZHANG, WEILI - CHEN, JIANJUN

Posters Index

Code	Title
0179	DEVELOPING OF COMPOSITE FROM BIOMASS-DERIVED ACTIVATED CARBONS WITH CARBON NANOTUBES FOR ELECTROCHEMICAL ELECTRODES TEMIRGALIYEVA, TOLGANAY - KUZUHARA, SOKI - NODA, SUGURU - NAZHIPKYZY, MERUYERT - MANSUROV, ZULKHAIR - LESBAYEV, BAKHYTZHAN - PRIKHODKO, NIKOLAY
0180	PREPARATION OF MESOCARBON MICROBEADS/SHORT CARBON FIBER COMPOSITES DERIVED FROM DIFFERENT SIZED MESOCARBON MICROBEADS IM, UISU - LEE, BYUNG-ROK - PECK, DONG-HYUN - JUNG, DOOHWAN
0184	NANOPOROUS CARBON COMPOSITES – SYNTHESES AND CHARACTERISTICS GAWDZIK, BARBARA - SOBIESIAK, MAGDALENA - PODKOŚCIELNA, BEATA - PODDUBNAYA, OLGA I. - PUZIY, ALEXANDER
0187	THEROMCHEMICAL CONVERSIONS OF DVB-MA-TMVS TERPOLYMERS DURING THEIR CARBONIZATION PROCESS SOBIESIAK, MAGDALENA - GAWDZIK, BARBARA - PODKOŚCIELNA, BEATA - PODDUBNAYA, OLGA I. - PUZIY, ALEXANDER
0196	IMPACT OF PRIMARY AND SECONDARY BEADS OF CARBON ENTEROSORBENT ON EPR PARAMETERS OF RATS WITH CYTOSTATIC NEUTROPENIA SNEZHKOVA, ELISAVETA - PYATCHANINA, TETIANA - BURLAKA, ANATOLIY - NIKOLAEV, VOLODYMYR - MAZUR, MARIA - KOZYNCHENKO, OLEKSANDR - SHEVCHUK, OKSANA
0199	SHRINKAGE MINIMIZATION IN ORGANIC GELS OBTAINED BY MICROWAVE HEATING CANAL-RODRÍGUEZ, MARIA - VILLANUEVA, SARA F. - MONTES-MORAN, MIGUEL - MENÉNDEZ, J. ANGEL - ARENILLAS, ANA
0200	COMPARATIVE EVALUATION OF TWO FORMULATIONS OF ACTIVATED CARBON USED AS ENTEROSORBENTS IN THE TREATMENT OF EXPERIMENTAL LIVER KOZYNCHENKO, OLEKSANDR - SNEZHKOVA, ELISAVETA - T.V., DATSKO - NIKOLAEV, VOLODYMYR - SYDORENKO, OLEKSIY - BARDAKHIVSKA, KVITOSLAVA - MAZUR, MARIA - L.A, SAKHNO
0202	BIOPHYSICAL STUDY OF THE EFFECTIVENESS OF TWO FORMULATIONS OF NANO/MACROPOROUS CARBON ENTEROSORBENTS IN THE TREATMENT OF EXPE KOZYNCHENKO, OLEKSANDR - SARNATSKAYA, VERONIKA - PROKOPENKO, IGOR - YUSHKO, LARYSA - SIDORENKO, ALEXEI - MASLENNY, VITALY - NIKOLAEV, VLADIMIR
0207	ION SELECTIVE SUBNANOMETER SIZED NANOWINDOWS OF SINGLE WALL CARBON NANOHORNS TANIGAKI, NAOTO - MURATA, KATSUYUKI - HAYASHI, TAKUYA - KANEKO, KATSUMI
0211	G-C3N4/CARBON COMPOSITES FOR LITHIUM-SULFUR BATTERIES ZHANG, HAN - ZHAO, ZONGBIN - HOU, YANAN - ZHANG, ZHICHAO - WANG, XUZHEN - QIU, JIESHAN
0214	SIMULTANEOUS TAGGING AND ENHANCEMENT OF CARBON FIBER/EPOXY INTERFACE BY UPCONVERSION LUMINESCENT NANOPARTICLES LIU, DAWEI - LI, GANG - YANG, XIAOPING
0216	ADSORPTION OF CO2 ON N-ENRICHED POROUS CARBON DERIVED FROM CROSSLINKED POLYETHYLENEIMINE SPHERE CHEN, SHUIXIE - LIU, FENGLEI - SUN, PENG
0219	PRIMARY AND SECONDARY BEADS OF CARBON ENTEROSORBENT AND OXIDATIVE STRESS' INDICES IN MELPHALAN SINGLE INJECTION SNEZHKOVA, ELISAVETA - SHEVCHUK, OKSANA - MIKHAILENKO, VIKTOR - GLAVIN, ALEXEI - MAKOVETSKA, LYUDMILA - BARDAKHIVSKA, KVITOSLAVA - NIKOLAEV, VOLODYMYR - SARNATSKAYA, VERONIKA - KOZYNCHENKO, OLEKSANDR
0220	CHARPHITE PROJECT: COAL CHAR AS A SUBSTITUTING MATERIAL OF NATURAL GRAPHITE IN GREEN ENERGY TECHNOLOGIES SANTOS, ANA - BIALECKA, BARBARA - CALUS-MOSZKO, JOANNA - CRUCERU, MIHAI - FREIRE, CRISTINA - GUEDES, ALEXANDRA - LAZARO-MARTINEZ, JUAN - POPESCU, LUMINITA - PREDEANU, GEORGETA - VALENTIM, BRUNO - WAGNER, NICOLA
0224	PREPARATION OF NOVEL MATERIALS FOR ELECTRODE OF ENERGY DEVICES BY ELECTRO SPINNING AND NANOHYBRID OSHIDA, KYOICHI - MINAMISAWA, TAKUNORI - ANDO, AKINOBU - KOBAYASHI, NOZOMI - MISAWA, DAIKI - ITAYA, TOMOYUKI - OSAWA, KOZO - MURATA, MASAHIKO - MORIYAMA, MINORU - HATA, TOSHIMITSU
0227	SYNTHESIS OF BIO-COMPATIBLE GRAPHENE QUANTUM DOTS BY THERMAL DECOMPOSITION OF SILICON CARBIDE LEE, NA EUN - LEE, SANG YOON - LIM, HYUNG SAN - PARK, YANG JEONG - KIM, JUNG WOO - LEE, JU HYUK - CHO, SUNG OH - JEONG, JONG-MIN - JEONG, WON-IL - CHO, YONG-HOON
0229	THE EFFECT OF TWO FORMULATIONS OF NANO/MACROPOROUS CARBON ENTEROSORBENTS ON OXIDATIVE STRESS INDEXES OF EXPERIMENTAL ANIMALS SANATSKAYA, VERONIKA - MIKHALENKO, VIKTOR - SHEVCHUK, OKSANA - GLAVIN, ALEXEI - MAKOVETSKA, LYDMILA - SAKHNO, LARYSA - YUSHKO, LARYSA - KOZYNCHENKO, OLEKSANDR - NIKOLAEV, VLADIMIR
0235	NITROGEN-DOPED MESOPOROUS CARBONS PREPARED BY USING COPOLYMER TEMPLATES FOR ENHANCED ADSORPTION OF HEAVY METAL IONS YE, GANG - SONG, YANG
0238	MOS2 DECORATED HOLEY GRAPHENE BASED CORE-SHEATH FIBERS FOR SUPERCAPACITORS WITH HIGH ENERGY DENSITY AND RATE CAPABILITY WANG, CHAOJUN - ZHAI, SHENGLI - CHEN, YUAN

Code	Title
0241	PREPARATION OF CARBON FIBER USING ASH FREE COAL PYROLYSIS RESIDUE LEE, SEON HO - LEE, SONG MI - PECK, DONG-HYUN - LEE, BYUNG-ROK - SHUL, YONG GUN - JUNG, DOO-HWAN
0242	FLEXIBLE VOC SENSORS USING CONDUCTIVE POLYMERS AND CARBON FIBER ELECTRODES FOR APPLICATION TO TEXTILES KIM, TAEKYEONG - JEONG, EUIGYUNG - BAE, JINSEOK - LEE, JUNHEON - PARK, JIHOON - LEE, JAEWOONG
0243	ULTRAFAST HYDROTHERMAL ASSEMBLY OF NANOCARBON MICROFIBERS IN NEAR-CRITICAL WATER FOR 3D MICRO-SUPERCAPACITORS ZHAI, SHENGLI - CHEN, YUAN
0245	VERTICAL MOS2 NANOSHEETS/HYDROGENATED GRAPHENE HYBRID CATALYSTS FOR EFFICIENT AND STABLE HYDRO-GEN EVOLUTION REACTION TONG, XILI - HAN, XIUXIU - GUO, XIANGYUN
0246	INVESTIGATION ON CHEMICAL STRUCTURE EVOLUTION OF REDUCED GRAPHENE OXIDE UNDER SUPER-HIGH TEMPERATURE KONG, QINGQIANG - CHEN, CHENGMENG
0254	THE EFFECT OF CNTS FOR CFRTPS MADE WITH POLYAMIDE 6 SHIBATA, MASATAKA - IRISAWA, TOSHIHIRA - TANABE, YASUHIRO
0255	ELASTIC REARRANGEMENT OF STACKED NANOGRAPHENES UNDER HIGH PRESSURE COMPRESSION CHOTIMAH, NURUL - PUTRI, AUSTINA D - ONO, YUJI - HATTORI, YOSHIYUKI - WANG, SHUWEN - FUTAMURA, RYUSUKE - URITA, KOKI - VALLEJOS-BURGOS, FERNANDO - NEIMARK, ALEXANDER V - KANEKO, KATSUMI
0259	MACROPOROUS MONOLITHIC CARBON XEROGEL VIA TEMPLATE-FREE METHOD FROM RESORCINOL AND FORMALDE-HYDE CHOI, GUN - LEE, JAE YOUNG - YOON, TAE-HO
0260	CARBON FIBERS SUPPORTED NISE NANOWIRE ARRAYS AS EFFICIENT AND FLEXIBLE ELECTROCATALYSTS FOR THE OXYGEN EVOLU-TION REACTION GUO, XIANGYUN - TONG, XILI - GUO, XIAONING
0262	APPLICATION OF THE SI-B-C-O COATING ON POROUS CARBON BODIES FOR HIGH TEMPERATURE THERMAL PROTECTION LI, TONG-QI - ZHANG, DAH-AI - JIAO, XING-JIAN - FENG, ZHI-HAI
0263	ACTIVATED CARBON FROM GRAPE SEEDS UPON HYDROTHERMAL CARBONIZATION AND CHEMICAL ACTIVATION MANZANO, JAVIER - VILLAMIL, JOHN - F. MENA, ISMAEL - DIAZ, ELENA - F. MOHEDANO, ANGEL - RODRIGUEZ, JUAN J.
0267	HYDROPHOBIC CARBON MATERIALS AS BIPHASIC MICRORREACTOR FOR CATALYTIC APPLICATIONS IN AQUEOUS MEDIA CALVO-CASTAÑERA, FELICIANO - ALVAREZ-RODRIGUEZ, JESUS - CANDELA, NURIA - MAROTO-VALIENTE, ANGEL
0269	NOVEL HYBRID MATERIAL BASED ON BIMETAL OXYHYDROXIDES (FE:CE)-GAC FOR SUPERIOR AS(V) REMOVAL VENCES-ALVAREZ, ESMERALDA - RANGEL-MENDEZ, RENE
0272	FLEXIBLE AND ULTRASENSITIVE FABRICS-BASED CARBON COMPOSITES SENSOR FOR DETECTING HUMAN MOTIONS LIANG, JINGJING - ZHAO, ZONGBIN - TANG, YONGCHAO - ZHANG, HAN - WANG, XUZHEN - QIU, JIESHAN
0273	N-DOPED MESOPOROUS CARBON NANOSHEETS DERIVED FROM MOFS FOR HIGH-RATE SUPERCAPACITORS HOU, YANAN - ZHAO, ZONGBIN - ZHANG, HAN - WANG, XUZHEN - QIU, JIESHAN
0274	SYNTHESIS OF GRAPHENE-CNT HYBRID MATERIALS AND THEIR ELECTROCHEMICAL APPLICATIONS SONG, JUN HO - RYU, JAE SEUNG - WOO, SANG CHEOL - LEE, CHURL SEUNG - PARK, JI SUN - KIM, YOUNG KWON
0278	VARIABLE TEMPERATURE RAMAN SCATTERING ON MICROMECHANICS OF CARBON FIBER REINFORCED POLYIMIDE/CNTS COM-POSITES RAN, MIN - SONG, YUNJIA - WU, QILIN
0280	ON THE RELIABILITY OF SODIUM CO-INTERCALATION IN EXPANDED GRAPHITE PREPARED BY DIFFERENT METHODS AS ANODES FOR SODIUM-ION BAT CABELLO, MARTA - BAI, XUE - CHYRKA, TARAS - ORTIZ, GREGORIO - LAVELA, PEDRO - ALCANTARA, RICARDO - TIRADO, JOSE
0282	THE STUDY OF INTERFACIAL ADHESION OF CFRTP MADE WITH IN-SITE POLYMERIZABLE THERMOPLASTICS KOBAYASHI, SARASA - TANABE, YASUHIRO - IRISAWA, TOSHIHIRA
0286	HIERARCHICALLY MICRO/MESOPOROUS CARBONS OBTAINED FROM ZIF-8/GRAPHENE OXIDE HYBRIDS BY A NOVEL NANOPORE LITHOGRAPHY APPROACH MARTIN-JIMENO, FRANCISCO JULIAN - SUAREZ GARCIA, FABIAN - PAREDES, JUAN IGNACIO - MARTINEZ-ALONSO, AMELIA - TAS-CÓN, JUAN M.D.
0294	SYNTHESIS AND HYDROGEN STORAGE PROPERTIES OF NICKEL NANOPARTICLES-CARBONACEOUS MATERIAL COM-POSITE KITAZAKI, AKIHITO - OHTSUKA, YOHEI - KINUMOTO, TARO - TSUMURA, TOMOKI - TOYODA, MASAHIRO

Posters Index

Code	Title
0297	A STUDY ON MECHANICAL STRENGTH AND DAMPING BEHAVIOR OF CARBON FIBERS/MODIFIED CNT-REINFORCED EPOXY COMPOSITE SHAFT CHOI, WOONGKI - KUK, YUNSU - YOON, GEUMRAN - CHOI, SUN-WOONG - PARK, YONG-MIN - SEO, MIN-KANG
0299	PRESSURIZED PHYSICAL ACTIVATION: EFFICIENT AND FACILE PRODUCTION METHOD OF HIGHLY POROUS ACTIVATED CARBONS YI, HYEONSEOK - NAKABAYASHI, KOJI - YOON, SEONG-HO - MIYAWAKI, JIN
0300	PREPARATION OF CHITIN DERIVED ACTIVATED CARBON AS AN EFFECTIVE REMOVAL ABSORBENT FOR FORMALDE-HYDE SICK-HOUSE GAS RYU, DONG-YEON - SHIMOHARA, TAKAAKI - NAKABAYASHI, KOJI - MIYAWAKI, JIN - YOON, SEONG-HO
0307	NANOTEXTURING TIO2 OVER CARBON NANOTUBES FOR HIGH ENERGY AND POWER DENSITY PSEUDOCAPACITORS IN ORGANIC ELECTROLYTES POIROT, NATHALIE - RAJALINGAM, VENKATESAN - RAYMUNDO-PIÑERO, ENCARNACION
0312	COAL BOTTOM ASH RECYCLING FOR RARE EARTH ELEMENTS RECOVERING FOR SOLAR COLLECTORS SANTOS, ANA - ABAGIU, ALEXANDRU - BIALECKA, BARBARA - CALUZ-MOSZKO, JOANNA - GONCALVES, PAULA - GUEDES, ALEXAN-DRA - POPESCU, LUMINITA - PREDEANU, GEORGETA - SLAVESCU, VALERICA - VALENTIM, BRUNO
0314	CARBON MATERIALS DOPED WITH AMINOCUMULENE AS DISPERSANT AND N-CONTAINING PRECURSOR ENHANCE CO2 ELECT-ROREDUCTION REACTION GERASIMOVA, ALENA - GARCÍA-CRUZ, LETICIA - MELEZHIK, ALEXANDER - ALEKHINA, OLGA - MONTIEL, VICENTE - TKACHEV, ALEK-SEY - INIESTA, JESÚS
0316	SYNTHESIS OF AMINOCUMULENE VIA POLYCONDENSATION OF HEXAMETHYLENETETRAMINE. SPECTRAL CHARACTERIZATION AND APPLICATIONS AS DISP GERASIMOVA, ALENA - GARCÍA-CRUZ, LETICIA - MELEZHIK, ALEXANDER - ALEKHINA, OLGA - DYACHKOVA, TATYANA - MONTIEL, VICENTE - TKACHEV, ALEKSEY - INIESTA, JESUS
0317	A STUDY OF ELECTRICAL, MECHANICAL AND THERMAL CHARACTERISTICS OF NON-WOVEN FABRIC HEATING SHEETS WITH CHOPPED CARBON FIBER LEE, SORA - KIM, WONSEOK
0320	HIGHLY ACTIVE NICKEL/CARBON CATALYSTS PREPARED BY CARBONIZATION OF A NICKEL/2-METHYL-IMIDAZOLE BASED ORGANOMETALLIC FRAMEWORK MARTÍN JIMENO, FRANCISCO JULIÁN - SUÁREZ GARCÍA, FABIÁN - PAREDES NACHÓN, JUAN IGNACIO - ALONSO MARTÍNEZ, AMELIA - DÍEZ TASCÓN, JUAN MANUEL
0327	POLYETHYLENE/CARBON NANOTUBES NANOCOMPOSITE FOR WATER VAPOR EXTRACTION FROM HUMID AIR ARAUJO PINHEIRO, ROMARIO - ARAUJO SILVA, AMANDA - TRAVA-AIROLDI, VLADIMIR JESUS - CORAT, EVALDO JOSE
0329	CARBON NANOTUBES UNZIPPED ON GRAPHITE AND THEIR CAPACITIVE PERFORMANCE IN SALT/BRACKISH WATER FOR CDI SYS-TEMS ARAUJO SILVA, AMANDA - ARAUJO PINHEIRO, ROMÁRIO - TRAVA-AIROLDI, VLADIMIR JESUS - CORAT, EVALDO JOSE
0330	APPLICATION OF CARBON NANOTUBES GROWN ON GRAPHITE FOR CAPACITIVE DEIONIZATION ARAUJO SILVA, AMANDA - ARAUJO PINHEIRO, ROMÁRIO - TRAVA-AIROLDI, VLADIMIR JESUS - CORAT, EVALDO JOSE
0337	HIGHLY PYRIDINIC N-DOPED CARBONS FROM ALGAE BIOMASS WITH HIGH PERFORMANCE TOWARDS SUPERCAPACITORS AND OXYGEN REDUCTION TIAN, ZHONGWEI - ZHANG, TIANYUN - CAI, JINJUN
0339	IMMERSION ENTHALPY OF ACTIVATED CARBONS WITH DIFFERENT OXYGEN CONTENT IN TOLUENE-HEXANE MIX-TURES HERNÁNDEZ-MONJE, DIANA - GIRALDO, LILIANA - MORENO-PIRAJAN, JUAN CARLOS
0340	GRAPHENE OXIDE MEMBRANES INTERCALATED WITH ZIRCONIUM-BASED METAL-ORGANIC FRAMEWORKS SUI, XIAO - CHEN, YUAN
0341	PREPARATION OF TIN OXIDE NANOFIBERS USING CARBON NANOFIBERS AS A TEMPLATE AND ITS APPLICATION TO SECONDARY BATTERY NAKABAYASHI, KOJI - YOSHIDA, SHO - MIYAWAKI, JIN - YOON, SEONG-HO
0343	DEVELOPING ENHANCED ANODE MATERIALS FOR LIB WITH HYPER COAL COATING ON THE GRAPHITE SURFACE NAKABAYASHI, KOJI - CHUNG, DABIN - MIYAWAKI, JIN - YOON, SEONG-HO
0344	ELECTROCHEMICAL PERFORMANCES OF JOULE-HEATED ACTIVATED CARBON FOR ENHANCED EDLC KIM, ICK-JUN - YANG, SUNHYE - IM, SANG-MIN
0347	INVESTIGATION OF REMOVAL MECHANISMS OF ACETALDEHYDE BY ANILINE SUPPORTED CELLULOSE-BASED ACTIVAT-ED CARBON FIBER RYU, DONG-YEON - SHIMOHARA, TAKAAKI - NAKABAYASHI, KOJI - UEDA, MORIO - KEIKO, IDETA - MIYAWAKI, JIN - MOCHIDA, ISAO - YOON, SEONG-HO
0350	BROMINATION OF ZIGZAG AND ARMCHAIR EDGES HIRAMATSU, SHINTA - MASAKI, SHIORI - YAMADA, YASUHIRO - KUBO, SHINGO - SATO, SATOSHI

Code	Title
0351	INFRARED-RAY SPECTROSCOPIC ANALYSIS OF BROMINATED GRAPHENE NANORIBBON HIRAMATSU, SHINTA - MASAKI, SHIORI - YAMADA, YASUHIRO - SATO, SATOSHI
0353	STRUCTURAL CONTROL OF NITROGEN-DOPED CARBON MATERIALS WITH THE AID OF COMPUTATIONAL CHEMISTRY MURATA, MASATOSHI - YAMADA, YASUHIRO - KUBO, SHINGO - SATO, SATOSHI
0356	SYNTHESIS OF MESOPOROUS CARBON WITH TURBOSTRATIC CARBON LAYER WALLS THROUGH IRON CITRATE AND ITS ELECTROCHEMICAL PROPERTY KITAZAKI, AKIHITO - KUROKI, CHIKA - TSIMURA, TOMOKI - KINUMOTO, TARO - TOYODA, MASAHIRO
0361	RAMAN SPECTROSCOPY ON SOOT PRODUCED FROM A MINI-CAST SOOT GENERATOR: IMPACT ON SOOT STRUCTURE FROM HEATING ... LE, KIM CUONG - T[illegible]R[illegible]K, SANDRA - PINO, THOMAS - BENGTSSON, PER-ERIK
0362	THE INFLUENCE OF HIGH-ACTIVE CARBON-CONTAINING MIXTURES ON COMBUSTION OF CONDENSED SYSTEMS BAKKARA, AYAGOZ
0367	PREPARATION OF $NIFE_2O_4$/MULTI-WALLED CARBON NANOTUBES HYBRIDS: APPLICATION FOR THE REMOVAL OF DI-CLOFENAC IN CWPO PROCESS HUACCALLO-AGUILAR, YSABEL - ALVAREZ-TORRELLAS, SILVIA - AGUEDA, V. ISMAEL - DELGADO, JOSE ANTONIO - OVEJERO, GABRI-EL - GARCIA, JUAN
0368	ALCOHOL DEHYDROGENATION ON KRAFT LIGNIN-DERIVED CHARS WITH SURFACE BASICITY ROSAS, JUANA Mª - GARCIA-MATEOS, FRANCISCO J - MOULEFERA, IMANE - BENYOUCEF, ABDELGHANI - RODRIGUEZ-MIRASOL, JOSE - CORDERO, TOMAS
0369	FLEXIBLE AND LOW-COST BINDERLESS CAPACITORS BASED ON P- AND N-CONTAINING FIBROUS ACTIVATED CARBONS FROM DENIM CLOTH WASTES CORDERO-LANZAC, TOMAS - GARCIA-MATEOS, FRANCISCO J - ROSAS, JUANA Mª - RODRIGUEZ-MIRASOL, JOSE - CORDERO, TOMAS
0371	ACTIVATED CARBONS AS CATALYTIC SUPPORT FOR CU NANOPARTICLES MARTIN, F. JAVIER - YANG, CHIA.MIN - GARCÍA MATEOS, FRANCISCO JOSÉ - GUERRERO-PÉREZ, M. OLGA - RODRIGUEZ-MIRASOL, JOSÉ - CORDERO, TOMÁS
0372	INFLUENCE OF MAJOR POST-COMBUSTION FLUE GAS SECONDARY COMPONENTS ON CO_2 ADSORPTION OVER RE-NEWABLE CARBON BASED MATERIALS ROSAS, JUANA Mª - CALVO-MUÑOZ, ELISA M. - GARCIA-MATEOS, FRANCISCO J - RODRIGUEZ-MIRASOL, JOSE - CORDERO, TOMAS
0379	ADSORPTION OF CO_2, CO, AND N_2 ON POROUS CARBONS WITH DIFFERENT MICROTEXTURAL AND SURFACE CHEMI-CAL HETEROGENEITIES KIM, MOON HYEON - YANG, KI HYUCK - NGUYEN, THI PHUNG THAO
0382	SYNTHESIS BY METHOD OF ELECTROSPINING OF COMPOSITE CARBON NANOFIBERS OF SPECIAL PURPOSE PRIKHODKO, NIKOLAY - SMAGULOVA, GAUKHAR - RAKHIMZHAN, NURGALY - LESBAYEV, BAKHYTZHAN - NAZHIPKYZY, MERUYERT - KAYDAR, BAYAN - MANSUROV, ZULKHAIR
0385	BEYOND KOH ACTIVATION FOR THE SYNTHESIS OF HIGHLY POROUS CARBONS FOR SUPERCAPACITOR ELECTRODES SEVILLA, MARTA - FERRERO, GUILLERMO A. - FUERTES, ANTONIO B
0386	TINIDAZOLE DEGRADATION ASSISTED BY SOLAR RADIATION AND IRON-DOPED SILICA XEROGELS ACOSTA-RANGEL, AURORA - SÁNCHEZ-POLO, MANUEL - SÁNCHEZ-POLO, ANA MARÍA - BERBER-MENDOZA, MARÍA SELENE - RIVE-RA-UTRILLA, JOSÉ - ROZALEN-ASTUDILLO, MARIA LUISA - LÓPEZ-RAMÓN, MARÍA VICTORIA
0389	SYNTHESIS OF RUTHENIUM CARBON NANOSPHERES AS CATALYST FOR ENVIRONMENTAL APPLICATIONS SERRA-PÉREZ, ESTRELLA - ÁLVAREZ-TORRELLAS, SILVIA - DELGADO, JOSÉ ANTONIO - AGUEDA, VICENTE ISMAEL - OVEJERO, GABRI-EL - GARCÍA, JUAN
0390	EFFICIENT REMOVAL OF BISPHENOL A BY ADSORPTION ONTO CARBON ADSORBENTS HENÁNDEZ-ABREU, ANA BELÉN - GOFFIN-MEDINA, SHAROL - NIETO-NARANJO, PATRICIA - ÁLVAREZ-TORRELLAS, SILVIA - ÁGUEDA, VICENTE ISMAEL - DELGADO, JOSE ANTONIO - OVEJERO, GABRIEL - GARCÍA, JUAN
0396	COAL CHAR AS ELECTROCATALYST FOR THE OXYGEN REDUCTION AND EVOLUTION REACTIONS NUNES, MARTA - SANTOS, ANA CLAUDIA - VALENTIM, BRUNO - FREIRE, CRISTINA
0397	HALIDE REMOVAL FROM AQUEOUS SOLUTION BY NOVEL SILVER-POLYMERIC MATERIALS SÁNCHEZ-POLO, ANA MARÍA - LÓPEZ-PEÑALVER, JESÚS J. - SÁNCHEZ-POLO, MANUEL - RIVERA-UTRILLA, JOSÉ - VELO-GALA, IN-MACULADA - LÓPEZ-RAMÓN, MARÍA VICTORIA
0399	SULFONIC ACID-FUNCTIONALIZED COAL FLY ASHES: NEW SOLID ACID CATALYSTS FOR ESTERIFICATION REACTIONS PEIXOTO, ANDREIA - SILVA, SONIA - NUNES, MARTA - SANTOS, ANA CLAUDIA - VALENTIM, BRUNO - FREIRE, CRISTINA
0400	USE OF NANOSTRUCTURED MATERIALS AS ANTICANCER DRUG CARRIERS ROZALEN-ASTUDILLO, MARÍA LUISA - ACOSTA-RANGEL, AURORA - SÁNCHEZ-POLO, ANA MARÍA - FERNÁNDEZ-PERALES, M. - SÁN-CHEZ-POLO, MANUEL - RIVERA-UTRILLA, JOSÉ - LÓPEZ-RAMÓN, MARÍA VICTORIA

Posters Index

Code	Title
0407	CARBON MONOLITHS BASED ON TWO CARBON PHASES AS ELECTRODES FOR SUPERCAPACITORS LILLO RODENAS, MARIA ANGELES - KUNOWSKY, MIRKO - MORENO-FERNÁNDEZ, MARIA ANGELES - ROJO, JOSE MARIA - IBAÑEZ, JOAQUÍN M.
0414	NEUTRON IRRADIATION EFFECTS ON NUCLEAR GRAPHITE MICROSTRUCTURE CONTESCU, CRISTIAN
0420	CATALYTIC GRAPHITIZATION OF NI-P COATED PAN-BASED CARBON FIBERS TZENG, SHINN-SHYONG - LIN, YU-HON - HUANG, YING-TZU
0430	DESIGN AND OPTIMISATION OF PEROVSKITE-CARBON COMPOSITE ELECTROCATALYSTS FOR ORR AND OER DOSZCZECZKO, SZYMON
0449	PREPARATION AND CHARACTERIZATION OF ACTIVATED CARBONS ISSUED FROM ALGERIAN COALS BENCHEIKH, WAHIBA - BELHACHEMI, MERIEME - ADDOUN, ABDELHAMID
0455	NOVEL KINETIC INTERPRETATION OF THE CARBON-NO REACTION: N2O VS. NO2 AS INTERMEDIATES. OYARZÚN-ARAVENA, ANDREA M.
0460	HIGHLY GRAPHITIZED CARBON NANOCAGES CONFINING SULFUR FOR HIGH PERFORMANCE LITHIUM-SULFUR BATTERIES XIA, YUAN
0461	MICROWAVE-ASSISTED SYNTHESIS OF A GRAPHENE– BI8LA10O27 NANOCOMPOSITE AS AN EFFICIENT CATALYTIC COUNTER ELECTRODE FOR DYE-SENS YUNRAPACH, AREEROB - OH, WON-CHUN
0462	DEFORMATION MECHANISMS OF PAN BASED-CNT COMPOSITES: MOLECULAR DYNAMICS (MD) SIMULATION APPROACH HEO, SO JEONG - CHAE, HAN GI - LEE, SEUNG GEOL
0463	INCREASE IN SPECIFIC SURFACE AREA AND CHANGE IN INTERNAL STRUCTURE OF CARBON BLACK IN RELATION TO ISOTHERMAL OXIDATION TIME LEE, SANG MIN - BAEK, UN GYEONG - ROH, JAE SEUNG
0464	THE EFFECT OF CARBON FILLERS IN BINDERS ON THE IMPROVED JOINING STRENGTH OF ISOTROPIC BULK GRAPHITE BAEK, UN-GYEONG - KANG, DONG-SU - LEE, SANG-MIN - ROH, JAE-SEUNG
0468	MODELING MULTISCALE KEROGEN POROSITY YANN, MAGNIN - ROLAND, PELLENQ - JEREMIE, BERTHONNEAU - FRANZ-JOSEPH, ULM
0475	A CARBON BLACK CEMENT COMPOSITE FOR TRAFFIC MONITORING COSTA, PEDRO - MONTEIRO, ANDRE - OESER, MARKUS - CACHIM, PAULO
0477	DISPERSION OF CARBON NANOTUBES IN POLYMER-MATRIX COMPOSITES INVESTIGATED BY THZ SPECTROSCOPY KANG, SEUNG BEOM - PARK, YOUNG SOO - KIM, JI YEON - KIM, WON-SEOK - CHOI, YOUNG CHUL - CHUNG, DONG CHUL
0483	PROTECTIVE EFFECT OF A CARBON AEROGEL IN METAL ORGANIC FRAMEWORK/CARBON AEROGEL SYSTEMS DOMÁN, ANDREA - MADARÁSZ, JÁNOS - LÁSZLÓ, KRISZTINA
0484	THE EFFECT OF DRY-BAG ISOSTATIC PRESSING ON COLD-STATE PERFORMANCE OF GRAPHITE MATRIX SPHERES LU, ZHENMING
0485	CRYSTALLINE CARBON NANOPARTICLES IN THERMORESPONSIVE SMART GELS BERKE, BARBARA - CZAKKEL, ORSOLYA - LÁSZLÓ, KRISZTINA
0496	HIERARCHICAL HEMATITE NANOSTRUCTURE FILM MODIFIED WITH CARBON-BASED MATERIAL FOR PHOTOCHEMICAL WATER SPLITTING GUO, QIAN
0499	CARBON MONOLITHS WITH HIERARCHICAL POROUS STRUCTURE FOR ALL-VANADIUM REDOX FLOW BATTERIES VIVO-VILCHES, JOSE F. - ETIENNE, MATHIEU - KARAKASHOV, BLAGOJ - CELZARD, ALAIN - FIERRO, VANESSA - BROSSE, NICOLAS - DUFOUR, ANTHONY
0503	CARBON NANOFIBERS AS ELECTRODES FOR VANADIUM REDOX COUPLES ELECTROCHEMISTRY VIVO-VILCHES, JOSE F. - ETIENNE, MATHIEU - CELZARD, ALAIN - FIERRO, VANESSA - BROSSE, NICOLAS - DUFOUR, ANTHONY
0506	SHAPED NANOPOROUS CARBON ADSORBENT FOR LOW-TEMPERATURE ADSORPTION ACCUMULATION OF NATURAL GAS SHKOLIN, ANDREY - FOMKIN, ANATOLY - MEN'SHCHIKOV, ILYA
0508	CREATION THE HEATING TEXTILE BY SYNTHESIS OF CARBON NANOTUBES ON THE GLASS CLOTH MANSUROV, NURKADYR - KIM, SOL - PRIKHODKO, NIKOLAY - SMAGULOVA, GAUKHAR
0512	MISCANTHUS BIOMASS WASTES AS PRECURSORS FOR ACTIVATED CARBONS GÓRSKA, BARBARA - GARCÍA RÍOS, AIDA - BABUCHOWSKA, PAULINA - PRZYGOCKI, PATRYK - BÉGUIN, FRANÇOIS

Code	Title
0518	THE CHARACTERISTICS OF HYDROGEN ADSORBED ON CARBONACEOUS MATERIALS. KUMAGAI, HARUO
0520	ELECTROCHEMICAL CAPACITORS BASED ON CARBON NANOSHEETS ASSEMBLED IN HOLLOW FIBERS SAGISAKA, KENTO - KIMURA, MUTSUMI - HATTORI, YOSHIYUKI
0528	SUPER-HIGH RATE SODIUM-IOM STORAGE PERFORMANCE OF BI2SE3/GRAPHENE LI, DAN - ZHOU, JISHENG - SONG, HUAIHE - CHEN, XIAOHONG
0529	A STUDY ON CFRP JOINT TEST ANALYSIS BY SURFACE TREATMENT LEE, MIN-GU - HUH, MONG-YOUNG - PARK, KIL-YOUNG
0530	CREATIVE SYNTHESIS OF N, O-CODOPED MESOPOROUS CARBON FOR SODIUM STORAGE ZHOU, CHUNLI - CHEN, XIAOHONG - JIA, MENGQIU - ZHOU, JISHENG - MA, ZHAOKUN - SONG, HUAIHE
0532	HETEROGENOUS FENTON REACTION USING CARBON-FE CATALYST COUPLED TO AN IN SITU H2O2 ELECTROGENERATION FOR WASTEWATER TREATMENT ZARATE-GUZMAN, ANA ITZEL - GONZALEZ-GUTIERREZ, LINDA VICTORIA - TORRES-DELGADO, GERARDO - ROMERO-CANO, LUIS ALBERTO - MEDEL-REYES, ALEJANDRO - CARRASCO-MARIN, FRANCISCO
0533	SUPERHYDROPHOBIC CARBON "LOTUS" LEAVES FABRICATED BY CARBON NANOTUBES WITH SUPERIOR LOADING CAPACITY JIA, YI - CHEN, FEIFEI - WEI, JINQUAN
0534	LAMINATE DESIGN FOR CFRP APPLIED TO THE INTERMEDIATE SHAFT YOON, SUNGWON - HA, JONGROK - CHO, JEHYOUNG - KIM, TAEYEOB
0535	COAL-DERIVED GRAPHENE QUANTUM DOTS PRODUCED BY ULTRASOUND EXFOLIATION AND THEIR CAPACITY FOR CU(II) DETECTION YATING, YATING - REN, SHAOZHAO - LI, KEKE - DANG, YONGQIANG - LIU, GUOYANG - FU, SHIQI - QIU, JIESHAN
0540	STRUCTURAL ANALYSIS OF ENCAPSULATED FULLERENES INTO CARBON NANORINGS TOJO, TOMOHIRO - INADA, RYOJI - SAKURAI, YOJI
0541	AN EVALUATION FOR THE PROCESS-ABILITY OF HIGH PERFORMANCE COMPOSITES USING LIQUID TRANSFER MOLDING HA, JONG-ROK - YOON, SUNG-WON - KIM, KYEONG-SOO - KIM, NAM-HUN - CHO, JE-HYOUNG
0543	STUDY ON GELATION BEHAVIORS OF PAN POLYMER ACCORDING TO WATER CONTENT IN PAN/DMSO SYSTEM KUK, YUNSU - SEO, MIN-KANG - KUK, YUN-SU - CHOI, WOONG-KI - YOON, GEUM-RAN
0544	ENERGY-INTENSIVE COMPOSITION FOR DELAYS GABDRASHOVA, SHOLPAN - TULEPOV, MARAT - DALTON, ALAN - BAISEITOV, DAUREN - RAKHOVA, NURGUL - KAZAKOV, YURI
0546	THE OXIDATIVE STABILIZATION OF ISOTROPIC PITCH FIBER UNDER PRESSURE NAKABAYASHI, KOJI - LIU, JINCHANG - SHIMANOE, HIROKI - MIYAWAKI, JIN - YOON, SEONG-HO
0547	DELAYING COMPOSITION BASED ON MODIFIED COMPONENTS GABDRASHOVA, SHOLPAN - TULEPOV, MARAT - DALTON, ALAN - RAKHOVA, NURGUL - MANSUROV, ZULKHAIR - YURI, KAZAKOV - IGOR, PUSTOVALOV - SPANOVA, GALIYA - HAMZINA, BOLASH
0549	PREPARATION OF MESOPHASE PITCH DERIVED FROM HYPER-COAL WITH HIGH YIELD NAKABAYASHI, KOJI - INOUE, TOSHINORI - HAMAGUCHI, MAKI - SHIMANOE, HIROKI - MIYAWAKI, JIN - YOON, SEONG-HO
0553	STRUCTURAL CONTROL OF NANOCARBON COMPLEX MURATA, MASATOSHI - SHIMANO, NORIMICHI - YAMADA, YASUHIRO - KUBO, SHINGO - SATO, SATOSHI
0555	FABRICATION OF COBALT@TITANIA COATED WITH NITROGEN-DOPED CARBON LAYERS AS AN ELECTROCATALYST FOR ORR MIYAKE, KOJI - TAKEMURA, TOSHIKI - ZHU, YEXIN - HIROTA, YUICHIRO - UCHIDA, YOSHIAKI - NISHIYAMA, NORIKAZU
0559	A NANO-TEMPLATING APPROACH FOR THE DEPOSITION OF CARBON INSIDE ZEOLITE Y BOONYOUNG, PAWAN - KASUKABE, TAKATOSHI - HOSHIKAWA, YASUTO - NISHIHARA, HIROTOMO - KYOTANI, TAKASHI - NUEANG-NORAJ, KHANIN
0562	GRAPHENE OXIDE PROPERTIES INFLUENCE MEMBRANE WATER FILTRATION MORELOS-GOMEZ, AARON - SÁNCHEZ-SALAS, ROQUE - CRUZ-SILVA, RODOLFO - ORTIZ-MEDINA, JOSUE - MUÑOZ-SANDOVAL, EMILIO - LÓPEZ-URÍAS, FLORENTINO - TAKEUCHI, KENJI - TERRONES, MAURICIO - ENDO, MORINOBU
0564	EDLC CHARACTERISTIC OF FINE CARBON FIBER PREPARED VIA ELECTROSPINNING OF COAL EXTRACT TOYODA, MASAHIRO - WATANABE, HIROKI - INOUE, TOSHINORI - SAKAMOTO, HISATOSHI - WADA, SHOHEI - HAMAGUCHI, MAKI - KIKUCHI, NAOKI - KINUMOTO, TARO - TSUMURA, TOMOKI
0566	DEVELOPMENT OF SULFUR-IMPREGNATED ACTIVATED CARBON FOR PHENOL ADSORPTION FROM PITCH LEE, SONG MI - LEE, SEON HO - LEE, BYUNG-ROK - PECK, DONG-HYUN - JUNG, DOO-HWAN

Code	Title
0569	A FACILE ONE-POT SYNTHESIS OF BI2S3/CARBON NANOFIBERS COMPOSITE AS AN ELECTRODE MATERIAL FOR SUPERCAPACITOR APPLICATIONS MOYSEOWICZ, AGATA
0577	TAILORING PROPERTIES OF REDUCED GRAPHENE OXIDE BY MEANS OF OXYGEN PLASMA KONDRATOWICZ, IZABELA - NADOLSKA, MAŁGORZATA - ŻELECHOWSKA, KAMILA - SAWCZAK, MIROSŁAW
0583	HIGHLY STRETCHABLE, SENSITIVE STRAIN SENSORS BASED ON COMPRESSED ANISOTROPIC GRAPHENE FOAM/POLYMER NANO-COMPOSITES ZENG, ZHIHUI - LU, XUEHONG
0586	OXIDIZED CARBON MATERIALS FOR THE HYDROLYSIS OF CELLULOSE AZAR*, FATIMA-ZAHRA - LILLO RÓDENAS, MARIA ÁNGELES - ROMÁN-MARTÍNEZ, MARI CARMEN
0590	BIOMASS CONVERSION TO HIGH ADDED VALUE PRODUCTS BY RU-CARBON CATALYSTS AZAR, FATIMA-ZAHRA - PETRYN, DMYTRO - LILLO RÓDENAS, MARIA ÁNGELES - ROMÁN-MARTÍNEZ*, MARI CARMEN
0595	PREPARATION AND APPLICATION OF VARIOUS PCNF COMPOSITES FOR LOW TEMPERATURE FUEL CELL LEE, SEON HO - LEE, SONG MI - LEE, BYUNG-ROK - PECK, DONG-HYUN - SHUL, YONG GUN - JUNG, DOO HWAN
0596	COMPOSITE MATERIAL BASED ON ORGANOMETALIC-DERIVED CARBON AND METAL OXIDE NANOPARTICLES WITH IMPROVED ELECTROCHEMICAL PERFORMA ARENAS ESTEBAN, DANIEL - ÁVILA BRANDE, DAVID - URONES GARROTE, ESTEBAN - GONZALEZ GARCIA, PEDRO - OTERO DIAZ, LUIS CARLOS
0598	NEW METHOD FOR HIGH FLUORINE CONTENT AND CRYSTALLINITY FLUORINATED GRAPHITE JUNG, MIN-JUNG - HAN, JEONG-IN - KIM, JUNG GU - LEE, YOUNG-SEAK
0600	GAS ADSORPTION PROPERTIES OF MESOPOROUS GRAPHENE-MOF COMPOSITES SZCZEŚNIAK, BARBARA - CHOMA, JERZY
0604	FABRICATION OF ALL-CARBON ELECTRIC FIELD TYPE ELECTRON EMISSION DEVICES USING METALLIC OR SEMICON-DUCTING SINGLE-WALLED CARBON KURODA, AKIHISA - SHIMOI, NORIHIRO - TOHJI, KAZUYUKI - SATO, YOSHINORI
0605	POLYPYRROLE-BASED MICROPOROUS CARBONS PREPARED IN THE PRESENCE OF ORGANIC POTASSIUM SALTS DZIURA, ALEKSANDRA - DZIURA, ROBERT - DRZEWIŃSKA, ANNA - CHOMA, JERZY
0608	HIGH-EFFICIENCY HEATING CHARACTERISTIC OF SIC FIBER MAT UNDER MICROWAVE CHO, KWANG YOUN - JOO, YOUNG JUN - KHISHIGBAYAR, KHOS-ERDENE
0609	PHOTOCATALYTIC OXIDATION OF PHENOL USING ZNO-DECORATED VERTICALLY ALIGNED CARBON NANOTUBES OLIVEIRA, INÊS - GIRÃO, ANA - SILVA, RICARDO - FARIA, JOAQUIM - SILVA, CLÁUDIA - SILVA, RUI
0610	ELECTROSTATIC PROPERTIES OF GRAPHENE NANORIBBONS UNDER AN EXTERNAL ELECTRIC FIELD GAO, YANLIN - OKADA, SUSUMU
0612	CROSSOVER FROM KAGOME TO DIRAC BANDS IN POROUS GRAPHITIC NETWORKS BY TOPOLOGICAL DESIGN MARUYAMA, MINA - OKADA, SUSUMU
0614	ELECTRONIC AND GEOMETRIC STRUCTURES OF GRAPHENE ADSORBING AL NANOPARTICLES UNDER AN EXTERNAL ELECTRIC FIELD MATSUBARA, MANAHO - OKADA, SUSUMU
0615	NOVEL CARBON-TI COMPOSITES BASED ON FUNCTIONALIZED CELLULOSE FOR PHOTOCATALYTIC WATER DECONTAM-INATION PEREZ-CADENAS, AGUSTIN FRANCISCO - HAMAD, HESHAM - BAILON-GARCIA, ESTHER - MORALES-TORRES, SERGIO - CARRAS-CO-MARIN, FRANCISCO - MALDONADO-HODAR, FRANCISCO JOSE
0616	PARTICLE SIZE REDUCTION AND CRYSTALLINITY CHANGE IN PAN-BASED CARBON FIBER LEE, SANG HYE - LEE, SANG MIN - HWANG, YOUNG MIN - ROH, JAE SEUNG
0618	SELF-TEMPLATED CARBON MATERIALS FROM THE CHLORINATION OF METAL ORGANIC FRAMEWORKS MARTIN-CALVO, SARA - ARENAS-ESTEBAN, DANIEL - SALCEDO-ABRAIRA, PABLO - HORCAJADA, PATRICIA - ÁVILA-BRANDE, DAVID
0620	THE USE OF P-PHENYLENEDIAMINE TO CONTROL GRAPHENE OXIDE MEMBRANE PORE-GAP AND STABILITY FOR WA-TER PURIFICATION KANDJOU, VEPIKA - HERNAEZ, MIGUEL - ACEVEDO, BEATRIZ - MAYES, ANDREW G. - MELENDI-ESPINA, SONIA - PEREZ MAS, ANA MATILDE
0625	TAILORING NITROGEN FUNCTIONAL GROUPS ON THE CARBON SURFACE FOR SUPERCAPACITOR ELECTRODE MATE-RIALS KIERZEK, KRZYSZTOF - CHOMIAK, KATARZYNA - MELENDI-ESPINA, SONIA - ACEVEDO, BEATRIZ - GONZÁLEZ, ZORAIDA - MOYSEO-WICZ, AGATA - MENÉNDEZ, ROSA - GRYGLEWICZ, GRAŻYNA
0629	MONOLITHIC GRAPHENE TREES AS ANODE MATERIAL FOR HYBRID CAPACITORS YANG, SUNHYE - JEONG, SEUNG YOL - IM, SANGMIN - KIM, MINCHUL - KIM, ICK-JUN

Code	Title
0630	SYNTHESIS OF GRAPHENE BY OXIDATION AND REDUCTION OF CU WITH ALCOHOL CHEMICAL VAPOR DEPOSITION OGATA, YUYA - GONNOKAMI, HIROMICHI - TACHIBANA, MASARU
0631	THERMAL PROPERTY OF EPOXY COMPOSITES WITH OXYFLUORINATED EXPANDED GRAPHITE CHOI, YE JI - LEE, KYEONG MIN - KANG, DA HEE - KONG, EUN YOUNG - LEE, YOUNG-SEAK
0633	GRAPHENE OXIDE MACROSCOPIC MULTILAYER ASSEMBLIES MASER, WOLFGANG - VÍCTOR-ROMÁN, SANDRA - HERNÁNDEZ-FERRER, JAVIER - ANSÓN-CASAOS, ALEJANDRO - GARCÍA-BORDEJÉ, ENRIQUE - BENITO, ANA M.
0636	FLUORINATION OF ACTIVATED CARBON FOR THE HARMFUL SUNSTANCE ADSORPTION PERFORMANCE IMPROVEMENT SONG, EUN JI - CHOI, YE JI - KIM, MIN-JI - JUNG, MIN-JUNG - LEE, YOUNG-SEAK
0638	PREPARATION OF NERVE AGENT SIMULANT GAS SENSOR USING POLYACRYLONITRLE BASED ACTIVATED CARBON FIBERS KANG, DA HEE - KIM, MIN-JI - CHOI, YE JI - RYU, SEUNG-KON - LEE, YOUNG-SEAK
0639	THE EFFECT OF FLUORINATION ON THERMAL AND MECHANICAL PROPERTIES OF CARBON NANOTUBE AND GRAPHENE NANOPLATELET REINFORCED EPOXY KONG, EUN YOUNG - KANG, DA HEE - LEE, KYEONG MIN - LEE, YOUNG-SEAK
0640	VOCS ADSORPTION CHARACTERISTICS OF CU INTRODUCED ACTIVATED CARBON FIBERS BY ELECTROLESS PLATING KIM, MIN-JI - KANG, DA HEE - CHOI, YE JI - KONG, EUN YOUNG - CHOI, SUK SOON - RYU, SEUNG-KON - LEE, YOUNG-SEAK
0641	GRAPHITE WITH HIGH RATE PERFORMANCE AS LIB ANODE BY INTRODUCING C-F BOND USING DIRECT-FLUORINATION KIM, KYUNG HOON - KANG, DA HEE - IM, JI SUN - LEE, YOUNG-SEAK
0642	VARIATION IN CARBON BLOCK DENSITY INFLUENCED BY ⊠-RESIN OF PETROLEUM-BASED BINDER PITCH KIM, JIWOOK - KIM, KYUNGHOON - SONG, EUNJI - LEE, YOUNG-SEAK
0643	EFFECT OF FLUORINATION AND CARBON-COATING ON SILICON AS LITHIUM-ION BATTERY ANODE HAN, JEON-IN - KIM, KYUNG HOON - JUNG, MIN-JUNG - LEE, YOUNG-SEAK
0646	ADTIVATED CARBONS OBTAINED FROM OLIVE WOOD AND DOPED WITH N AND S AS CATALYSTS FOR OXYGEN REDUCTION REACTION PEREZ-CADENAS, AGUSTIN FRANCISCO - CASTELO-QUIBEN, JESICA - MALDONADO-HODAR, FRANCISCO JOSE - CARRASCO-MARIN, FRANCISCO
0653	ONE-STEP SYNTHESIS OF NANOPOROUS CARBONS WITH ULTRAHIGH SURFACE AREA AND THEIR USE AS ELECTRODES FOR SUPER-CAPACITORS DIEZ, NOEL - SEVILLA, MARTA - FERRERO, GUILLERMO A. - FUERTES, ANTONIO B.
0655	GREEN GRAPHENE OXIDE REDUCTION ASSISTED BY SUPERCRITICAL WATER GASIFICATION OF GLYCEROL TORRES, DANIEL - PINILLA, JOSE LUIS - SUELVES, ISABEL
0656	SYNTHESIS AND INVESTIGATION OF CARBON SORBENTS BASED ON HIGH VOLATILE BARZAS COAL ISMAGILOV, ZINFER - KOZLOV, ALEXEI - ZYKOV, IGOR
0657	CONTROL OF THE SIZE OF HOLLOW CARBON SUBMICROSPHERES WITH POROUS SHELL HEINTEL, TOM - ALONSO-MORALES, NOELIA - MOYA, CRISTIAN - LEMUS, JESUS - SANTIAGO, RUBEN - PALOMAR, JOSE - GILARRANZ, MIGUEL A.
0671	SYNTHESIS OF NITROGEN-DOPED GRAPHENE OXIDES AS POTENTIAL MEMBRANE MATERIALS FOR BIOGAS UPGRADING KARAHAN, H. ENIS - GOH, KUNLI - BIRER, OZGUR - CHEN, YUAN - BAE, TAE-HYUN
0676	KRAFT LIGNIN UPGRADING IN SUPERCRITICAL WATER WITH NI-CEO2/CARBON BASED CATALYSTS: EFFECT OF THE CARBON SUPPORT CARDOSO, ADERLANIO - RAMÍREZ-REINA, TOMÁS - SUELVES, ISABEL - PINILLA, JOSÉ LUIS - MILLAN, MARCOS - HELLGARDT, KLAUS
0677	IN-DEPTH INVESTIGATION OF THE CLOSED POROSITY OF GLASSY CARBONS BY SCATTERING TECHNIQUES BADACZEWSKI, FELIX
0681	NANOGOLD AS ADDITIVE TO PSEUDOHALIDE ELECTROLYTES FOR CARBON/CARBON ELECTROCHEMICAL CAPACITOR BUJEWSKA, PAULINA - GORSKA, BARBARA - FIC, KRZYSZTOF
0689	ELECTRICAL CONDUCTION MECHANISM IN CARBON-CERAMIC COMPOSITES PEREZ-CADENAS, MARÍA - TRAUTWEIN, GUIDO - ALCAÑIZ-MONJE, JUAN
0691	AQUEOUS PHASE REFORMING OF BREWERY WASTEWATER USING CARBON SUPPORTED PT CATALYSTS OLIVEIRA, ADRIANA - BAEZA, JOSÉ ALBERTO - CALVO, LUISA - ALONSO-MORALES, NOELIA - HERAS, FRANCISCO - RODRÍGUEZ, JUAN JOSÉ - GILARRANZ, MIGUEL ANGEL
0695	KINETIC STUDIES OF MESOPHASE FORMATION IN PITCH EATON, DAVID - OWEN, AARON - WEISENBERGER, MATTHEW - ANDREWS, RODNEY

Posters Index

Code	Title
0698	SYNTHESIS OF MONOLITHIC CARBON MATERIALS WITH HIERARCHICALLY ORDERED PORE STRUCTURE AS MODEL SYSTEMS FOR ENERGY STORAGE APPLICATIONS RUMSWINKEL, SIMON - KOCZWARA, CHRISTIAN - ELSAESSER, MICHAEL S. - PARIS, OSKAR - HÜSING, NICOLA
0703	SCREENING OF DIFFERENT NI-NOBLE METAL BIMETALLIC CATALYSTS SUPPORTED ON CARBON NANOFIBERS FOR ONE-POT CELLOBIOSE CONVERSION FRECHA, ESTHER - PUEYO, ALVARO - TORRES, DANIEL - SUELVES, ISABEL - PINILLA, JOSÉ LUIS
0707	ON THE INFLUENCE OF THE CARBON SUPPORT IN THE HYDRODEOXYGENATION OF GUAIACOL OVER SUPPORTED MOLYBDENUM CARBIDE CATALYSTS OCHOA, ELBA - TORRES, DANIEL - AYLON, ELVIRA - PINILLA, JOSÉ LUIS - SUELVES, ISABEL
0709	COMPARATIVE STUDY OF HARMFUL ALGAL BLOOM BIOMASS-DERIVED BIOSORBENT, POROUS CARBON, AND FE/C COMPOSITE FOR CR (VI) REMOVAL CUI, YANBIN - ATKINSON, JOHN
0712	BORON DOPED DIAMOND THERMISTOR FOR IN-SITU DENTAL APPLICATIONS NETO, MIGUEL A. - SILVA, RUI - OLIVEIRA, FILIPE J.
0718	STABILITY OF CARBON-SUPPORTED PT CATALYSTS ON THE AQUEOUS PHASE REFORMING OF TUNA-COOKING WASTEWATER OLIVEIRA, ADRIANA - BAEZA, JOSE ALBERTO - CALVO, LUISA - ALONSO-MORALES, NOELIA - HERAS, FRANCISCO - LEMUS, JESUS - RODRIGUEZ, JUAN JOSE - GILARRANZ, MIGUEL ANGEL
0720	CO2 CAPTURE ONTO REDUCED GRAPHENE OXIDE (RGO) IMPREGNATED WITH POLYETHYLENIMINE (PEI) ALVAREZ-TORRELLAS, SILVIA - DELGADO, JOSÉ ANTONIO - UGUINA, Mª ÁNGELES - OVEJERO, GABRIEL - GARCÍA, JUAN - AGUEDA MATE, ISMAEL VICENTE - LARRIBA, MARCOS - PALOMO, JORGE - WENDELBO, RUNE
0721	GRAPHENE OXIDE MEMBRANES IN OCULAR REGENERATIVE MEDICINE GONZALEZ, ZORAIDA - ZAMBRANO-ANDAZOL, IRIANA - BLANCO, CLARA - MENENDEZ, ROSA - MEANA, ALVARO - VAZQUEZ, NATALIA - CHACON, MANUEL - SANCHEZ-AVILA, RONALD M. - SIERRA, MARIA - FERNANDEZ-VEGA, ALVARO - SANCHEZ, TERESA - MERAYO-LLOVES, JESUS
0722	INFLUENCE OF NANOSTRUCTURE AND SURFACE CHEMISTRY OF CARBON SUPPORTS IN THE PERFORMANCE OF PD-CU/C NITRATE REDUCTION IN WATER ALONSO-MORALES, NOELIA - BAEZA, JOSE A. - HERAS, FRANCISCO - CALVO, LUISA - RODRIGUEZ, JUAN J. - GILARRANZ, MIGUEL A.
0725	SELF-PROPAGATING HIGH TEMPERATURE SYNTHESIS OF REFRACTORY BORON CARBIDE POWDER SEIDUALIEVA, AIJAN - KAMUNUR, KASTER - MANSUROV, ZULHAIR
0736	ASSISTED DEVELOPMENT OF MESOPHASE IN PITCH WITH THE USE OF DISPERSED GRAPHENE WEISENBERGER, MATTHEW - OWEN, AARON - EATON, DAVE - ANDREWS, RODNEY
0737	THE PRACTICAL FUNDAMENTALS OF AN ADSORBENT FOR ADSORBED NATURAL GAS VEHICLES HOLBROOK, BILLY-PAUL - BURNS, NIKOLAS - ONUBOGU, KEN - REED, BRAD - WILLIAMS, ROGER
0738	DEACTIVATION OF ZIRCONIUM-PHOSPHORUS CARBON CATALYST ON THE DEHYDRATION OF METHANOL TO DME REACTION TORRES-LIÑÁN, JAVIER - GARCÍA-ROLLÁN, MIGUEL - PALOMO, JOSE - ROSAS, JUANA MARIA - RODRÍGUEZ-MIRASOL, JOSE - CORDE-RO, TOMÁS
0740	GRAPHENE PRODUCTION BY A COMBINATION OF MECHANICAL WET MILLING AND ULTRASOUND MANCILLAS, SERGIO - LÓPEZ, EDDIE
0747	SYNTHESIS AND CHARACTERIZATION OF MESOSTRUCTURED HYBRID CARBON NANOFIBER/REDUCED GRAPHENE OXIDE JASSO-RAMOS, LUIS - OJEDA-HERNANDEZ, ALFREDO - GUERRERO-BERMEA, CYNTHIA - GARCIA-GOMEZ, NORA - MANRIQUEZ, JUAN - CRUZ-SILVA, RODOLFO - SEPULVEDA-GUZMAN, SELENE
0748	BORON-NITROGEN HETERODOPING IN MULTIWALLED CARBON NANOTUBES OLVERA MARTINEZ, JOSE A - MUÑOZ SANDOVAL, EMILIO - LÓPEZ URIAS, FLORENTINO
0749	SILVER STRUCTURES ANCHORED TO NITROGEN-DOPED CARBON NANOTUBE SPONGES: EFFECT OF NITROGEN FUNCTIONALITIES QUINTERO GONZÁLEZ, CLAUDIA C. - LOPEZ URIAS, FLORENTINO - MUÑOZ SANDOVAL, EMILIO
0751	POST-MORTEM AC CHARACTERIZATION AS THE INSIGHT INTO AGEING PROCESS OF ELECTROCHEMICAL CAPACITORS PIWEK, JUSTYNA - FIC, KRZYSZTOF - PLATEK, ANETTA - GHIMBEU, CAMELIA - FRACKOWIAK, ELZBIETA
0752	ACETANILIDE AND ANTIPYRINE ADSORPTION ON LIGNIN-DERIVED ACTIVATED CARBON GUTIÉRREZ, MARIA DEL CARMEN - GARCÍA-MATEOS, FRANCISCO JOSÉ - MARTÍN-CORRAL, MONICA - RODRÍGUEZ-MIRASOL, JOSÉ - CORDERO, TOMÁS
0754	MULTIMODAL POROUS CARBON DERIVED FROM IONIC LIQUIDS: CORRELATION BETWEEN PORE SIZES AND IONIC CLUSTERS JEONG, JUN HUI - SHIN, DALWOO - KIM, KWANG BUM

Posters Index

Code	Title
0759	HIGH TEMPERATURTE TREATMENTS OF POROUS ACTIVATED CARBON GARCÍA-MATEOS, FRANCISCO JOSÉ - RUIZ-ROSAS, RAMIRO - ROSAS, JUANA Mª - RODRÍGUEZ-MIRASOL, JOSÉ - CORDERO, TOMÁS
0760	SALT-TEMPLATED CARBONS AS ELECTRODE MATERIAL IN AQUEOUS ELECTROCHEMICAL CAPACITORS PLATEK, ANETTA - FIC, KRZYSZTOF - GHIMBEU, CAMELIA - NITA, CRISTINA - FRACKOWIAK, ELZBIETA
0761	SYNTHESIS AND ELECTRICAL CHARACTERIZATION OF HYBRID NANOCARBON FILMS OF REDUCED GRAPHENE OXIDE (RGO) AND GRAPHENE QUANTUM DOT NOVOA, ILSE - JHONY, JACOB - VAZQUEZ, SOFIA - GARCIA, NORA - SHAJI, SADASIVAN - SEPULVEDA, SELENE
0763	ALCOHOL DECOMPOSITION ON BASIC/ACID LIGNIN-DERIVED SUBMICRON DIAMETER CARBON FIBERS GARCIA-MATEOS, FRANCISCO JOSE - ROSAS, JUANA Mª - RODRÍGUEZ-MIRASOL, JOSE - CORDERO, TOMÁS
0766	DOPED CARBON NANOWALLS FOR POWER SOURCES EVLASHIN, STANISLAV - MIRONOVICH, KIRILL - DAGESYAN, SARKIS - DYAKONOV, PAVEL - ZIRYANOV, SERGEY - ZOTOVICH, ALEXEY - VAVILOV, SERGEI - AKHATOV, ISKANDER - SUETIN, NIKOLAY
0767	EVALUATION OF DIFFERENT PRECURSORS CONTAINING NITROGEN FOR THE CARBON NANOTUBE SPONGES PRODUCTION RODRIGUEZ CORVERA, CRISTINA DE LOURDES - LÓPEZ URÍAS, FLORENTINO - MUÑOZ SANDOVAL, EMILIO
0771	TRANSFORMATION OF TANNING INDUSTRY LEATHER SHAVINGS INTO CR-FUNCTIONALIZED CARBON MATERIALS ARCIBAR, JAVIER - BARAJAS, BERTHA SILVIA - CABELLERO, FELIPE - RANGEL, RENE
0775	BALL-MILLED SILICON CARBIDE TO FABRICATED NITROGEN-DOPED CARBON NANOTUBES. LAMAS MARTÍNEZ, KARLA JOHANA - LÓPEZ URIAS, FLORENTINO - MUÑOZ SANDOVAL, EMILIO - RODRÍGUEZ VARGAS, ISAAC
0782	CORRELATION OF RHEOLOGICAL PARAMETERS OF ACTIVATED CARBON COATINGS WITH THEIR ELECTROCHEMICAL RESPONSE IN AQUEOUS ELECTROLYTE GALEK, PRZEMYSLAW - FIC, KRZYSZTOF - FRACKOWIAK, ELZBIETA
0783	GLYCEROL-BASED ACTIVATED CARBON MICROSPHERES FOR THE REMOVAL OF AN IODINATED X-RAY CONTRAST MEDIA BATISTA, MARY - MESTRE, ANA S. - FONSECA, ISABEL
0785	RHEOLOGICAL PROPERTIES OF BIO-PITCHES: INFLUENCE OF VACUUM DISTILLATION PARAMETERS ON FLOW PROPERTIES KILIC, MURAT - ÖZSIN, GAMZENUR - APAYDIN-VAROL, ESIN
0786	SULFUR EFFECT ON THE GROWTH OF NITROGEN-DOPED NANOTUBE SPONGES BY CHEMICAL VAPOR DEPOSITION METHOD HERRERA HARO, LUIS ARMANDO - MUÑOZ-SANDOVAL, EMILIO - LOPEZ URÍAS, FLORENTINO
0787	ELECTROCHEMICAL PERFORMANCE TO CHARGE TRANSFER OF CUO-DECORATED CARBON NANOFIBERS RODRÍGUEZ-OLGUÍN, MIGUEL A. - SEPÚLVEDA-GUZMÁN, SELENE - DE LA PARRA-ARCINIEGA, SALOMÉ M. - SÁNCHEZ, EDUARDO M. - GARZA-MONTES-DE-OCA, NELSON FEDERICO - GARCÍA-GÓMEZ, NORA A.
0788	SUPERACTIVATED CARBON OBTAINED FROM POLYANYLINE CARBONIZED BY PHYSICAL ACTIVATION RUIZ-ROSAS, RAMIRO - TRAUTWEIN, GUIDO - ALCAÑIZ-MONGE, JUAN - PEREZ-CADENAS, MARIA
0795	ELECTROCHEMICAL OXIDATION OF ORDERED MESOPOROUS CARBONS: INFLUENCE OF CARBON GRAPHITIZATION. PÉREZ-RODRÍGUEZ, SARA - SEBASTIÁN, DAVID - GARCÍÁ, ANA BEATRIZ - LÁZARO, Mª JESÚS
0796	ACTIVATED ION-TRACK CARBON FOR FLEXIBLE SUPERCAPACITOR ELECTRODES LAUSEVIC, PETAR - PEJOVIC, PREDRAG - NIKOLIC, VLADIMIR - LAUSEVIC, ZORAN - TOIMIL-MOLARES, MARIA EUGENIA
0798	PLASMA IN LIQUID PROCESSING OF NANOMATERIALS SUSPENSION VIZIREANU, SORIN - IONITA, MARIA DANIELA - IONITA, EUSEBIU ROSINI - STOICA, SILVIU DANIEL - CARPEN, LAVINIA - TEODORESCU, MAXIMILIAN - ION, IOANA - DINESCU, GHEORGHE
0803	MANGANESE OXIDE DOPING ACTIVATED CARBON AND ITS APPLICATION AS ELECTRODE FOR ELECTROCHEMICAL SUPERCAPACITORS ELMOUWAHIDI, ABDELHAKIM - BAILON-GARCIA, ESTHER - CASTELLO-QUIBEN, JESICA - PEREZ-CADENAS, AGUSTIN F. - CARRASCO-MARIN, FRANCISCO
0804	NITROGEN DOPED CARBON SPHERES AS ELECTRODE MATERIAL FOR SUPERCAPACITORS ELMOUWAHIDI, ABDELHAKIM - BAILON-GARCIA, ESTHER - PEREZ-CADENAS, AGUSTIN F. - FIERRO, VANESSA - CELZARD, ALAIN - CARRASCO-MARÍN, FRANCISCO
0806	ADVANCED MICROPOROUS ACTIVATED CARBONS OBTAINED ON A BASIS OF RICE HUSK PAVLENKO, VLADIMIR - ABBAS, QAMAR - ISMAGILOV, ZINFER - MANSUROV, ZULKHAIR - SUPIYEVA, ZHAZIRA - PRZYGOCKI, PATRYK - BÉGUIN, FRANÇOIS
0812	PREPARATION OF A CARBON MONOLITH REPLICATING THE NANOPOROUS STRUCTURE AND SHAPE OF NANOPOROUS GLASS HOSHIKAWA, YASUTO - MASHIYAMA, TAKAHIRO - TANABE, KAORI - FUJIWARA, MITSUTERU - ITO, MUTSUHIRO - KYOTANI, TAKASHI
0813	REACTIVITY OF GRAPHENE OXIDE TOWARDS POLYAMINES VERNET-GARCÍA, JOAN - LÓPEZ-ROSA, CRISTINA - LÓPEZ-DE LA TORRE, MARÍA DOLORES - MELGUIZO, MANUEL - LÓPEZ-GARZÓN, JAVIER - GARCÍA-GALLARÍN, CELESTE

Posters Index

Code	Title
0815	SOLVOTHERMAL SYNTHESIS METHOD FOR THE PREPARATION OF O, P AND S DOPED GRAPHITIC CARBON NITRIDES MARTINEZ CARTAGENA, MANUEL EDUARDO - FERNANDEZ TAVIZON, SALVADOR - ROMERO GARCIA, JORGE - TORRES LUBIAN, JOSE ROMAN
0818	CHARACTERIZATION OF CARBON-CARBON COMPOSITES MURDIE, NEIL - TANG, LONGGUI
0820	COMPRESSIBILITY CHANGES BY THE ADSORPTION OF NITROGEN DOPED NANOTUBES WITH LIPID LANGMUIR MONOLAYERS GUERRA-CORREA, AXEL - GARCIA-GONZALEZ, ALCIONE - MUÑOZ-SANDOVAL, EMILIO - RUIZ-GARCIA, JAIME - VALDES, DONATO - LOPEZ-URIAS, FLORENTINO - GARCIA-REYES, BERNARDO
0821	PHYSICAL PROPERTIES OF SYNTHETIC DISTORTED NANOGRAPHENES INCORPORATING SEVEN-MEMBERED CARBOCYCLES PALOMINO-RUIZ, LUCÍA - MILLÁN, ALBA - GONZÁLEZ, M. TERESA - MARCOS, M. LUISA - CRUZ, CARLOS M. - MARQUEZ, IRENE R. - FUENTES, NOELIA - G. CAMPAÑA, ARACELI
0822	SYNTHESIS OF NA-X ZEOLITE FROM COAL ASSOCIATED OIL SHALE ASH BY ALKALINE FUSION HYDROTHERMAL METHOD BAI, SHU XIA BAI - CHU, CHU MO
0825	TURNING TAR-DERIVED PHENOLIC OILS INTO POROUS CARBON GELS FOR WATER POLLUTANTS REMOVAL SCHAEFER, SÉBASTIEN - GADONNEIX, PHILIPPE - SANCHEZ-SANCHEZ, ANGELA - CELZARD, ALAIN - FIERRO, VANESSA
0826	THERMALLY-DRIVEN HYDROGEN COMPRESSION BASED ON ADSORPTION ONTO ACTIVATED CARBONS SDANGHI, GIUSEPPE - SCHAEFER, SÉBASTIEN - NICOLAS, VINCENT - MARANZANA, GAEL - CELZARD, ALAIN - FIERRO, VANESSA
0830	ORDERED MESOPOROUS CARBONS OBTAINED BY SOFT-TEMPLATING OF TANNIN IN MILD CONDITIONS SANCHEZ-SANCHEZ, ANGELA - IZQUIERDO, MARÍA TERESA - MEDJAHDI, GHOUTI - GHANBAJA, JAAFAR - CELZARD, ALAIN - FIERRO, VANESSA
0831	STUDY OF CARBONS FROM GRAPEFRUIT PEELS FUNCTIONALIZED WITH OXYGEN AND NITROGEN AS ANODES FOR ION-NA BATTERIES ROMERO-CANO, LUIS A. - GARCÍA-ROSERO, HELENA - GONZÁLEZ-GUTIÉRREZ, LINDA V. - RAMOS-SÁNCHEZ, GUADALUPE - CARRASCO-MARIN, FRANCISCO
0832	SYNTHESIS OF CARBON GELS BY SOL-GEL POLYMERIZATION OF COAL-DERIVED PHENOLIC OILS WITH FORMALDEHYDE SANCHEZ-SANCHEZ, ANGELA - IZQUIERDO, MARÍA TERESA - SCHAEFFER, SEBASTIEN - CELZARD, ALAIN - FIERRO, VANESSA
0833	ELECTROCHEMICAL PERFORMANCES OF KOH-ACTIVATED, TANNIN-BASED, ORDERED MESOPOROUS CARBONS CASTRO-GUTIÉRREZ, JIMENA - SANCHEZ-SANCHEZ, ANGELA - IZQUIERDO, MARÍA TERESA - CELZARD, ALAIN - FIERRO, VANESSA
0834	HYDROLYSABLE TANNINS AS ALTERNATIVE PRECURSORS TO SYNTHESISE ORDERED MESOPOROUS CARBONS MONSIVAIS-ROCHA, JOSÉ EDUARDO - CASTRO-GUTIÉRREZ, JIMENA - SANCHEZ-SANCHEZ, ANGELA - CELZARD, ALAIN - FIERRO, VANESSA
0836	IN VITRO ADSORPTION STUDIES OF DICLOFENAC REMOVAL BY RICE HUSK DERIVED CHARCOALS ABDULKARIMOVA, ROZA - JAKPAR, JANDOSOV - BERNARDO, MARIA - VÉLEZ VÉLEZ, FÉLIX - FONSECA, ISABEL - SILVESTRE-ALBERO, JOAQUIN - LYUBCHYK, SVITLANA - MANSUROV, ZULKHAIR - MIKHALOVSKY, SERGEY
0839	IN SITU LFA THERMAL DIFFUSIVITY OF STABILIZED PAN FIBER AS A FUNCTION OF CARBONIZATION TEMPERATURE BURGESS, JORDAN - CRADDOCK, JOHN - WEISENBERGER, MATTHEW
0840	INFLUENCE OF THE DISPERSION METHOD OF GRAPHENE MATERIALS ON THE TRANSPORT PROPERTIES OF EPOXY BASED NANOCOMPOSITES SANCHEZ-HIDALGO, RUBEN - RALLINI, MARCO - YUSTE-SANCHEZ, VANESA - GONZALEZ, ZORAIDA - BLANCO, CLARA - VERDEJO, RAQUEL - TORRE, LUIGI - LOPEZ-MANCHADO, MIGUEL ANGEL - MENENDEZ, ROSA
0844	VERTICALLY ALIGNED LITHIOPHILIC CUO NANOSHEETS ON A CU COLLECTOR TO STABILIZE LITHIUM DEPOSITION FOR LITHIUM METAL BATTERIES ZHANG, CHEN - LV, WEI - HUANG, ZHIJIA - ZHANG, YUNBO - KANG, FEIYU - YANG, QUAN-HONG
0847	MODIFYING PROCESSING OF A MEXICAN COAL TO PREPARE FOUNDRY COKE PÉREZ-AGUILAR, NANCY VERÓNICA - GÓMEZ-GUTIÉRREZ, ANTONIO HUMBERTO - HERNÁNDEZ-CHAVERO, JESÚS EDGARDO - BARRIOCANAL-RUEDA, CARMEN - RODRÍGUEZ-MONREAL, AURORA - ARCIBAR-OROZCO, JAVIER ANTONIO
0852	PRODUCTION OF CARBON-CERAMIC SORBENTS AND FILTERS FOR WATER PURIFICATION KERIMKULOVA, ALMAGUL - AZAT, SEITKHAN - TOPANOV, BOLAT - KERIMKULOVA, MAKPAL - MANSUROV, ZULKHAIR
0853	A BINDER-FREE CARBON ELECTRODE SURFACE-MODIFIED BY RUTHENIUM OXIDE FOR LITHIUM-OXYGEN BATTERIES SHIN, HYUN-SEOP - JUNG, KYUN-NAM - HWANG, JIN-HA - LEE, JONG-WON
0854	ADSORPTION CAPACITY OF DIFFERENT COAL AND BIOMASS ACTIVATED CARBONS PREPARED BY A NEW CLEAN TECHNOLOGY PREDEANU, GEORGETA
0858	EVALUATION OF STRUCTURAL CHANGES IN ACTIVATED CARBONS AFTER NONANE ADSORPTION MARTINEZ ESCANDELL, MANUEL - FERREIRA DE PAULA, FABIANO - MONTERO LAGO, ROCHEL

Code	Title
0861	NOVEL COUNTER ELECTRODES BASED ON FEW-LAYER GRAPHENE OXIDE AND GRAPHENE OXIDE QUANTUM DOTS FOR DYE-SENSITIZED SOLAR CELLS TORRES, DANIEL - SEBASTIAN, DAVID - PINILLA, JOSE LUIS - SUELVES, ISABEL - LAZARO, MARIA JESUS - ARICÒ, ANTONINO S. - BAGLIO, VINCENZO
0866	PREPARATION OF HIGH OXIDATION RESISTANCE CARBON-CERAMIC COMPOSITES USING SELF-SINTERING MATERIALS OLAYA-GALIPIENSO, LUCIA - FUENTES PEREZ, JENIFER - SERNA ESPINOSA, MANUEL - SILVESTRE ALBERO, JOAQUIN - MARTINEZ ESCANDELL, MANUEL
0867	HIGH CO2 STORAGE CAPACITY ACTIVATED CARBON PREPARED FORM POLYURETHANE FOAM WASTES CAMPELLO-GOMEZ, IGNACIO - SILVESTRE, JOAQUIN - MARTINEZ-ESCANDELL, MANUEL
0868	NITROGEN DOPED CARBON MATERIALS AS PT-RU SUPPORT FOR THE ELECTRO-OXIDATION OF METHANOL SEBASTIAN, DAVID - NIETO-MONGE, MARIA JESUS - PASTOR, ELENA - LAZARO, MARIA JESUS
0871	ALLOTROPIC CARBON FORMS TO IMPROVE THE PERFORMANCE OF OPTOELECTRONIC ORGANIC DEVICES GOMEZ-AGUILAR, RAMON - RAMOS-DIAZ, ANTONIO ROWLAND - MARTINEZ-GUTIERREZ, HUGO - ORTEGA-CERVANTEZ, GERARDO
0872	OXIDATIVE DESULFURIZATION OF DIESEL FUEL CATALYZED BY CARBON-SUPPORTED POLYOXOMETALATES NUTTALL, CHARLOTTE - ROGERS, CORAL - HODGKISS, SOPHIE - CRAVEN, MICHAEL - KOZHEVNIKOVA, ELENA - KOZHEVNIKOV, IVAN
0877	STRUCTURE AND ELECTRICAL PROPERTIES OF HEAVILY NITROGEN DOPED CARBON ZHEREBTSOV, DMITRY - SMOLYAKOVA, KSENIA - YANTSEN, RUSLANA - ZHIVULIN, DMITRY - ZHIVULIN, VLADIMIR - EREMYASHEV, VIACHESLAV - HSU, HUA-SHU - GUO, FENG-WEI - ZAKHARCHUK, IVAN - LAHDERANTA, ERKKI
0878	GRAPHITIZING CARBONACEOUS MATERIALS USING AN ATOMIC ABSORPTION SPECTROMETER PUTMAN, KATE - SOFIANOS, VERONICA - ROWLES, MATTHEW - HARRIS, PETER - BUCKLEY, CRAIG - MARKS, NIGEL - SUAREZ-MARTINEZ, IRENE
0880	STUDY OF THE EFFECT OF COMPOSITIONS OF SURFACE-ACTIVE SUBSTANCES ON HIGH-VISCOSITY OIL SALAKHOV, RENAT - . KHAMIDULLIN, RENAT - BODIKOV, DYISEN - SEITZHANOVA, MAKPAL
0886	STUDY OF INFLUENCE OF THE ELECTROHYDRAULIC EFFECT ON HIGH-VISCOSITY OIL BODYKOV, DUISSENBAY - ABDIKARIMOV, MEIRAMGALI - SEITZHANOVA, MAKPAL - MANSUROV, ZYLKAIR
0887	TURNING LOW-VALUE COAL TAR DERIVED LIQUIDS INTO CARBONACEOUS MATERIALS AS ANODES FOR LI-ION BATTERIES CASTRO-MUÑIZ, ALBERTO - LÉONARD, ALEXANDRE F. - LORENZO-FIERRO, SARA - MARTÍNEZ-ALONSO, AMELIA - TASCÓN, JUAN M.D. - SUÁREZ-GARCÍA, FABIÁN - JOB, NATHALIE - PAREDES, JUAN I.
0890	REDUCED GRAPHENE OXIDE AS SOURCE-DRAIN ELECTRODES TO OTFT DEVICES RAMOS-DÍAZ, ANTONIO ROWLAND - GÓMEZ-AGUILAR, RAMON - ORTÍZ-LÓPEZ, JAIME - MARTÍNEZ-GUTIÉRREZ, HUGO
0897	PAN NANOFIBER MEMBRANE PREPARED VIA MULTI-NEEDLE ELECTROSPINNING AND NEEDLELESS ELECTROSPINNING FAN, YALING - LIU, JIE - WANG, CHUNHUA - LIANG, JIEYING - WANG, XIAOXU
0898	STUDY ON THE EXOTHERMIC BEHAVIOR OF LARGE-TOW PAN BASED CARBON FIBERS DURING THERMAL STABILIZATION GUO, LICHUANG GUO - LIANG, JIEYING LIANG - WANG, CHUNHUA WANG - LIU, JIE LIU - WANG, XIAOXU WANG
0899	STUDY ON EUTECTIC SALTS AS HIGH-EFFICIENCY CATALYSTS TO PREPARE HIGH-PERFORMANCE POLYACRYLONITRILE-BASED ACTIVATED CARBON FIB SANG, HUAYU - MIAO, PENG - LIU, JIE - WANG, XIAOXU
0901	HIGH PERFORMANCE FE-DOPED MN/TIO2 ACTIVATED CARBON CATALYST APPLIED IN PHOTO-FENTON-PEROXONE PROCESSES SECULA, MARIUS SEBASTIAN - CAGNON, BENOÎT - CHEDEVILLE, OLIVIER
0902	ENHANCED HIGH POWER PROPERTY ON ACTIVATED CARBON BASED CAPACITOR JUNG, MIN-JUNG - KO, YOONYOUNG - KIM, HONG GUN - RYU, SEUNG KON - LEE, YOUNG-SEAK
0908	PREPARATION AND ELECTRICAL PROPERTIES OF SOLID POLYMER ELECTROLYTES CONTAINING NEW THIOL-ENE CROSSLINKERS IN LIB HAN, MIJEONG - SUK, JUNGDON - KWAK, HYUNJO O - KANG, YONGKU
0909	RESEARCH ON THE PHOTOCATALYTIC ACTIVITY OF TIO2-COATED PAN-BASED ACTIVATED CARBON FIBER TIAN, YANG - AN, YONGZHEN - LIU, JIE - WANG, XIAOXU
0912	FREESTANDING SULFUR-GRAPHENE OXIDE/CARBON COMPOSITE PAPER AS A STABLE CATHODE FOR HIGH PERFORMANCE LITHIUM SULFUR BATTERIES KIM, JIM MIN - SUK, JUNGDON
0913	MODELING AND MEASUREMENT OF THREE-DIMENSIONAL TEXTURE OF CARBON MATERIALS BY OPTICAL MICROSCOPY ANDO, AKINOBU - OSHIDA, KYOICHI - TAKAHARA, JUN - TAKAHIDE, KIMURA - SASAKI, RYO
0915	THERMAL BEHAVIOR OF PAN PRECURSOR UNDER INERT ATMOSPHERE HAO, WEIZHE - ZHANG*, XUEJUN - TIAN, YANHONG

Posters Index

Code	Title
0916	THERMAL BEHAVIOR OF PBO PRECURSOR DURING LOW-TEMPERATURE CARBONIZATION HAO, WEIZHE - ZHANG*, XUEJUN - TIAN, YANHONG
0917	PREPARATION AND PROPERTIES OF FE3O4 / GRAPHENE MODIFIED WATERBORNE POLYIMIDE SIZING AGENT FOR CARBON FIBER HAO, RUITING - ZHANG*, XUEJUN - TIAN, YANHONG
0926	SYNTHESIS OF NANOHYBRID CARBON STRUCTURES-HORSERADISH PEROXIDASE (ARMORACIA RUSTICANA) (CNDS-HRP) FOR PHENOL DETECTION VALCÁRCEL MENDEZ, SANDRA VIVIANA - QUINTERO SAUMETH, JORGE - CASTILLO LEÓN, JOHN JAIRO - MEJIA OSPINO, ENRIQUE - CABANZO HERNÁNDEZ, RAFAEL
0929	ANALYSIS OF SURFACE RESISTANCE OF CARBON FIBER SURFACE HEATING ELEMENT ACCORDING TO HIGH CELLU-LOSE THICKENER CONTENT JEONG, HUN - LEE, MIN SANG - BACK, SEONG HUN - YUN, SEONG MIN - KO, SANG CEHOL - KWAC, LEE KU
0931	COMPARATIVE MECHANICAL AND THERMAL PROPERTIES OF CFRP ACCORDING TO COMPRESSION TEST GO, SUN HO - KIM, YONG JIN - TUGIRUMUBANO, ALEXANDRE - SHIN, HEE JAE - KANG, SUNG SOO - KIM, HONG GUN
0939	PREPARATIONANDCHARACTERIZATIONOFNANOCOMPOSITE OF RGO@TIO2 SUPPORTEDON FIQUE FIBERS FORPHOTOCATALYT-ICDEGRADATIONOFORGANICSDYES BARAJAS RUEDA, YEIMI LORENA - PINZÓN MARTÍNEZ, MARTHA LIZETH - GUTIERREZ NIÑO, NELSÓN - DIAZ RAMÍREZ, GERMÁN - MEJÍA OSPINO, ENRIQUE - CABANZO HERNÁNDEZ, RAFAEL
0940	SYNTHESIS AND CHARACTERIZATION OF POLYURETHANE FOAM OBTAINED FROM REDUCED GRAPHENE OXIDE AND 4,4'-MDI GUTIERREZ NIÑO, NELSON - MEJIA OSPINO, ENRIQUE - CABANZO HERNÁNDEZ, RAFAEL
0941	ENHANCED EFFICIENCY OF THERMAL STABILIZATION USING ELECTRON BEAM IRRADIATION FOR THE PREPARATION OF PAN-BASED CARBON FIBERS LIU, YUCHEN - WANG, XIAOXU - LIU, JIE - LIANG, JIEYING
0943	THE INFLUENCE OF INACCESSIBLE POROSITY IN THE ELECTROACTIVITY OF EXPANDED GRAPHITE/GEOBACTER SUL-FURREDUCENS BIOELECTRODES RAMÍREZ-MORENO, MARINA - ORTIZ, JUAN MANUEL - BERENGUER, RAUL - ESTEVE-NÚÑEZ, ABRAHAM
0945	INFLUENCE OF THERMAL-RESISTANT TEMPLATE ON SIZE AND FLUORESCENCE PROPERTIES OF CARBON DOTS MIKHRALIIEVA, ALBINA - BOTELHO DA MOTTA, HENRIQUE - ZAITSEV, VOLODYMYR
0948	FUNCTIONALIZATION OF CARBON NANOTUBES WITH DRUGS IN ORDER TO INCREASE THE BIOCOMPATIBILITY MARIANA, PRODANA - MARIUS, ENACHESCU - ANDREI BOGDAN, STOIAN - MIHAELA, BALAS - MADALINA-ANDREEA, BADEA - ANCA, DINISCHIOTU - CAMELIA, GAVAN
0949	APPLICATION OF MULTIWALLED CARBON NANOTUBES AS DRUG CARRIER SYSTEMS DANIELA, IONITA - SORIN MIRCEA, OPRESCU - BIANCA, TINCU - MIHAELA, BALAS - MADALINA ANDREEA, BADEA - ANCA, DINISCHIO-TU
0954	SYNERGISTIC EFFECT OF THE COMBINED USE OF GRAPHENE OXIDE AND MWCNT IN THE LONG-TERM STABILITY OF STARCH BIONANOCOMPOSITES MONTALBÁN, MERCEDES G. - DOMENE, DANIEL - GARCÍA-QUESADA, JUAN C. - MARTÍN-GULLÓN, IGNACIO
0956	GRAPHENE OXIDE AND MWCNT AS NANOREINFORCEMENTS OF STARCH-BASED BIONANOCOMPOSITES PLASTICIZED BY GLYCEROL AND BY IONIC LIQUIDS MONTALBÁN, MERCEDES G. - DOMENE, DANIEL - GARCÍA-QUESADA, JUAN C. - VÍLLORA, GLORIA - MARTÍN-GULLÓN, IGNACIO
0962	A NEW TYPE OF OXYGEN EVOLUTION CATALYST COMBINED WITH CARBON BASED MATERIALS KIM, HYUNAH - KANG, KISUK
0963	INFLUENCE OF CARBON NATURE IN THE HYDRODECHLORINATION OF CHLOROFORM TO VALUABLE HYDROCARBONS BY PD/C CATALYSTS FERNANDEZ, CARLOS - ANDREOLI, SARA - BEDIA, JORGE - WINCEK, RONALD - RODRIGUEZ, JUAN J. - GÓMEZ-SAINERO, LUISA MA-RIA - ESER, SEMIH
0967	EFFECT OF PRETREATMENT ON THE CHARACTERISTICS AND THERMAL PROPERTIES OF CARBON NANOFLUIDS FOR USE IN HEATING/COOLING SYSTEMS UYSAL, NILPERI - APAYDIN-VAROL, ESIN
0971	EVALUATION OF WASTE TIRES BY THERMAL TREATMENTS FOR APPLICATIONS IN ELECTRICAL DEVICES GONZÁLEZ GONZÁLEZ, REYNA BERENICE - MENDOZA DOMÍNGUEZ, ALBERTO - CABALLERO MATA, PORFIRIO
0975	INVESTIGATION ON THE ADSORPTION QUALITY OF BIOCHAR FOR WASTE WATER TREATMENT BY CARBONIZED BIO-MASS WASTE AND COAL SLIMES TOSUN, YILDIRIM
0977	CARBON NANOPOROUS MATERIALS WITH A HIGH HEALING EFFECT BEXEITOVA, KALAMPIR - NURALY, ASSIYA - AKNAZAROV, SESTAGER - PAVLYUKOV, ANDREY
0978	STUDY OF THE USE OF NATURAL MICROFERTILIZER PHYTOMICROFERTILIZER. ZHUMAKHAN, KAIRAT - AMZEEVA, ULPAN - SINYAVSKY, YURI

Code	Title
0980	CARBON FIBER MANUFACTURING CONDITIONS TO IMPROVE INTERFACIAL ADHESION STOJCEVSKI, FILIP
0981	KINETICS AND MECHANISM OF CATALYTIC COKE GASIFICATION BY AIR AND BY CO2 LEMOS, AMELIA - CARABINEIRO, SONIA - BOBADILLA, LUIS - CENTENO, MIGUEL - LEMOS, FRANCISCO - LOBO, LUIS
0982	CARBIDES AND GRAPHENE BASED MATERIALS FOR ELECTROCHEMICAL PRODUCTION OF HYDROGEN MONTERO, CÉSAR - DÍAZ-COELLO, SERGIO - RODRÍGUEZ, JOSÉ LUIS - ARÉVALO, CARMEN - GARCÍA, GONZALO - PASTOR, ELENA
0987	COMBUSTION OF WOOD CHARCOAL IN A CONICAL SPOUTED BED COMBUSTOR SAN JOSE, MARIA J - ALVAREZ, SONIA - LOPEZ, RAQUEL
0990	VALORISATION OF TEAK WOOD WASTES FOR PRODUCTION OF ACTIVATED CARBONS CANSADO, ISABEL - MOURÃO, PAULO
0991	FLUORESCENT GOLD CARBON DOTS FOR DETECTION OF MERCURY GOVINDARAJU, SARAVANAN - KIM, JONGSUNG - YUN, KYUSIK
0992	FLEXIBLE C-MO2C FIBER FILM WITH SELF-FUSED JUNCTIONS ZHANG, WENJIE - LV, RUITAO - KANG, FEIYU
0993	MESOPOROUS CARBON HOLLOW MICROSPHERES WITH RED BLOOD CELL LIKE MORPHOLOGY FOR EFFICIENT MICROWAVE ABSORPTION AT ELEVATED TEMP XU, HAILONG - YIN, XIAOWEI - ZHANG, LITONG - CHENG, LAIFEI
0994	IN SITU GROWTH OF ONE-DIMENSIONAL SIC NANOWIRES IN POROUS SC2SI2O7 CERAMICS WITH EXCELLENT MICROWAVE ABSORPTION PROPERTIES WEI, HANJUN - YIN, XIAOWEI - ZHANG, LITONG - CHENG, LAIFEI
0995	ELECTROMAGNETIC WAVE ABSORPTION PROPERTY OF RGO COMPOSITES HOU, ZEXIN
0996	ENHANCE THE SENSITIVITY OF CARBON NANOTUBES AS A NANOSENSER ALI, SHAMS - HORROCKS, BENJAMIN - HOULTON, ANDREW
0997	MODIFICATION OF MULTI-WALLED CARBON NANOTUBES BY QUADRUPLE HYDROGEN BONDING TO IMPROVE THE ELECTRICAL PROPERTIES OF POLYAMIDE HAN, M - LIM, HYUNWHA
0998	PREDICTION OF OXIDATIVE STABILIZATION BASED ON THE REACTIVITY OF PITCH FRACTIONS GUAN, TAOTAO - LI, YONGXIN - ZHANG, GUOLI - ZHAO, JIANGHONG - WANG, JIANLONG - LI, KAIXI
1001	CAGING NONCARBONS IN PRECISELY DESIGNED GRAPHENE NETWORKS FOR SUPERIOR VOLUMETRIC LITHIUM STORAGE HAN, JUNWEI - KONG, DEBIN - WEI, LV - TANG, DAI-MING - YANG, QUAN-HONG
1002	RECENT DEVELOPMENT OF FABRICATING FLEXIBLE MICRO-SUPERCAPACITORS FOR WEARABLE DEVICES LIU, YUQING - BEIRNE, STEPHEN - SMITH, EMILY - RAZAL, JOSELITO - CREAN, CAROL - WALLCE, GORDON - CHEN, JUN
1003	FABRICATION OF BIOMASS-BASED ANTI-MICROBIAL CARBON COMPOSITES FOR APPLICATION TO WOUND DRESSING KWON, OH HYEONG - KO, YOUNG-GWANG - CHO, DONGHWAN - PARK, WON HO
1004	ACTIVATED ELECTROSPUN CARBON NANOFIBERS FOR CARBON DIOXIDE ADSORPTION CHIANG, YU-CHUN - CHEN, YU-JEN - WU, CHENG-YEN
1005	HETEROGENEOUS SEEDED GROWTH OF ELECTRODEPOSITED LITHIUM METAL ON SINGLE-LAYERPOLYCRYSTALLINE GRAPHENE ZHANG, HAO - ZHANG, WENFENG - CAO, GAOPING
1007	LITHIUM ION BATTERY PERFORMANCES OF PHOSPHORUS-ENCAPSULATED CARBON NANOTUBE ANODES WITH DIFFERENT AMOUNTS OF PHOSPHORUS TOJO, TOMOHIRO - YAMAGUCHI, SHINPEI - AOYANAGI, KENGO - INADA, RYOJI - SAKURAI, YOJI
1008	INSIGHT INTO CONTROLLABILITY AND PREDICTABILITY OF PORE STRUCTURES IN PITCH-BASED ACTIVATED CARBONS LI, KAIXI - GUAN, TAOTAO - ZHANG, GUOLI - ZHAO, JIANGHONG
1009	A COMPARATIVE STUDY OF THE INTERACTION OF TWO DIFFERENT IRON PHTALOCYANINES WITH CARBON NANOTUBES PÉREZ-CADENAS, MARÍA - ASEDEGBEGA-NIETO, ESTHER - CARTER, JONATHAN - ANDERSON, JAMES A. - RODRÍGUEZ-RAMOS, INMACULADA - GUERRERO-RUIZ, ANTONIO R.
1010	MICROSTRUCTURE AND MECHANICAL PROPERTIES OF CARBON FIBER AND SIC NANOWIRES REINFORCED PYC MATRIX COMPOSITES LIN, HONGJIAO - LI, HEJUN
1011	SIMULTANEOUSLY IMPROVING THE MECHANICAL STRENGTH AND ELECTROMAGNETIC INTERFERENCE SHIELDING OF CARBON/CARBON COMPOSITES BY EL LI, HEJUN

A MOLECULAR DYNAMICS STUDY ON THE TRIBOLOGICAL PROPERTIES OF DIAMOND-LIKE CARBON IN HYDROGEN GAS ENVIRONMENT

Yang Wang[1*], Jingxiang Xu[1], Yusuke Ootani[1], Yuji Higuchi[1], Nobuki Ozawa[1], and Momoji Kubo[1]

[1] *Institute for Materials Research, Tohoku University, 2-1-1 Katahira, Aoba-ku, Sendai 980-8577, Japan*

*Presenting author's e-mail: yang.wang@imr.tohoku.ac.jp

Introduction

Diamond-like carbon (DLC), an amorphous carbon consisting of sp^3- and sp^2-hybridized carbon atoms, is widely used as a promising protective coating in many applications such as engine and bearings due to the good mechanical properties. It is reported that DLC coatings show the excellent lubricity in the hydrogen gas environment.[1] Erdemir et al. pointed out that the excellent lubricity of the DLC coatings is caused by the tribochemical reactions of DLC with hydrogen gas.[2] Moreover, hydrogen gas not only results in the low friction coefficient, but also reduces the wear of the DLC coatings.[3] To improve the tribological properties of the DLC coatings, it is necessary to reveal the low friction and low wear mechanisms induced by the tribochemical reactions between hydrogen gas and DLC; nevertheless, these understandings are still lacking by experiments due to the complex atomic-scale friction processes and tribochemical reactions at the friction interface. To deal with these difficulties, we use a molecular dynamics method to investigate the detailed tribochemical reaction dynamics between DLC and hydrogen gas and clarify the origin of the low friction and low wear properties of DLC.

Simulation models

We use our developed simulator, "LASKYO" based on the reactive force field (ReaxFF)[4], to reveal the tribochemical reaction dynamics. To model the contact of asperities at the real friction interface, we construct two half cylindrical hydrogen-terminated DLC models with a radius of 5.0 nm and pack them together with an overlap of 1.0 nm as shown in Figure 1. In the friction simulation, the top layer atoms of the upper DLC asperity are rigid and are slid to x-direction with 100 m/s. The bottom layer atoms of the lower DLC asperity are fixed during the friction simulation. Thermostat is applied to 0.5 nm layers next to the rigid layers by using the Berendsen thermostat. The system temperature is set at 300 K. The velocity-verlet algorithm is used to calculate the motion of each atoms with a time step of 0.25 fs/step under the periodic boundary conditions. To clarify the effect of hydrogen gas on the tribological behaviours of DLC, friction simulations are performed in both the vacuum and hydrogen gas environment for totally 600 ps. To simulate the hydrogen gas, hydrogen molecules are put around DLC asperities. Hydrogen gas pressure of 500 atm is used to accelerate the reactions between hydrogen molecules and DLC.

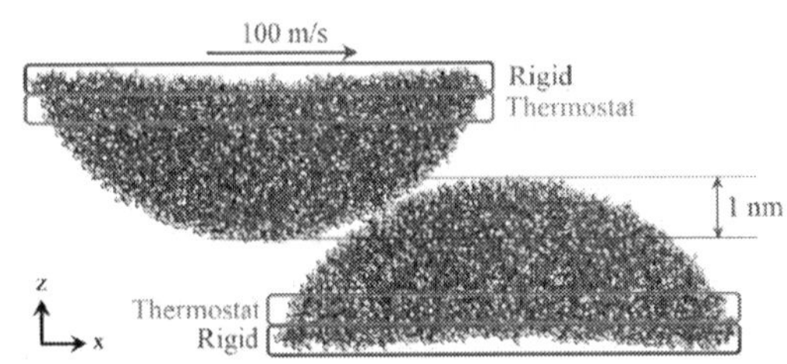

Figure 1. Simulation model of DLC asperities. Grey and white indicate the C and H atom, respectively.

Results and Discussion

In both vacuum and hydrogen gas environments, DLC asperities are collided together and separated away cyclically during the friction. Furthermore, in both simulations, we observe the generation of gaseous hydrocarbons such as methane, ethane, and ethylene during the friction. These gaseous hydrocarbons are desorbed from the surface, resulting in a depletion of the surface hydrogen terminations. For the friction behaviours of DLC in the vacuum, the depletion of hydrogen terminations causes the formation of interfacial C-C bonds connecting two DLC surfaces. Thus, two DLC asperities are strongly adhered together due to these interfacial C-C bonds, finally inducing an increase in the friction force and lose of the good lubricity. However, in the hydrogen gas environment, we observe less interfacial C-C bonds between the two DLC surfaces. This reduction of interfacial C-C bonds is due to the chemical reactions between DLC and hydrogen molecules as shown in Figure 2. In the vacuum, the number of hydrogen terminations decreases during friction, indicating that hydrogen terminations are depleted gradually. On the other hand, in the hydrogen gas environment, the number of hydrogen terminations holds almost no change and the number of hydrogen molecules decreases progressively during friction, indicating the passivation of hydrogen molecules to DLC surfaces. Thus, the depleted hydrogen terminations are replenished by reacting with the hydrogen molecules, finally inhibiting the interfacial bond formation and maintaining the low friction properties. For the wear behaviours, two wear mechanisms of DLC in our simulations are found: 1) chemical wear as the generation and desorption of gaseous hydrocarbons from DLC surface and 2) adhesive wear as the transfer of carbon atoms to the counter surface when the adhered DLC asperities are separated away. Figure 3 shows the detailed amount of (a) chemical and (b) adhesive wear. The amount of adhesive wear is much higher than that of chemical wear in both vacuum and hydrogen gas environments, indicating that the adhesive wear is the dominant wear mechanism for DLC. Furthermore, the amount of chemical wear in both simulations show the similar degree, whereas the amount of adhesive wear in the hydrogen gas environment is much lower than that in the vacuum. The adhesive wear of DLC is largely suppressed by the hydrogen gas due to the inhibition of interfacial bond formation between two DLC surfaces, resulting in a low wear behaviours.

Conclusions

In this study, we use MD simulations to study the atomic-scale friction process of DLC coatings in the vacuum and hydrogen gas environment. We find that chemical reactions between hydrogen gas and DLC surfaces replenish the depleted surface terminations during friction, maintaining the low friction behaviours and lowering the adhesive wear of the DLC coatings. Our simulation results successfully reveal the origin of the low friction and wear properties of the DLC coatings in the hydrogen gas environment.

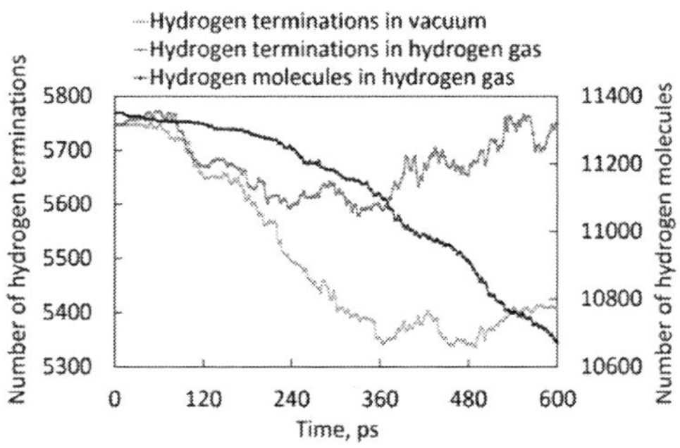

Figure 2. The number of hydrogen terminations and hydrogen molecules.

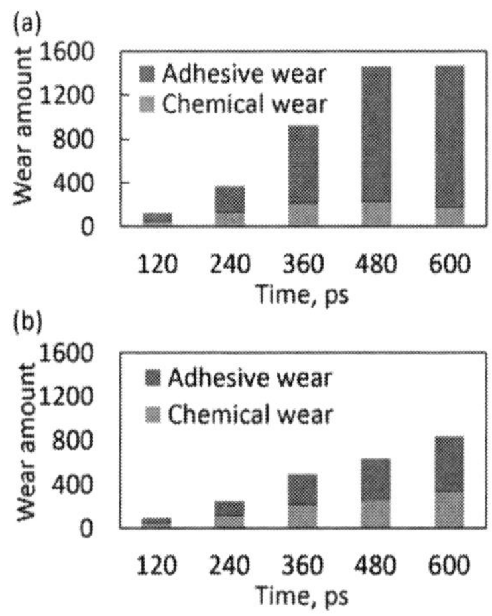

Figure 3. Wear amount of DLC asperities in the (a) vacuum and (b) hydrogen gas.

References

1. Okubo, H.; Tsuboi, R.; Sasaki, S. Frictional Properties of DLC Films in Low-Pressure Hydrogen Conditions. *Wear* 2015, 340–341, 2–8.
2. Erdemir, A.; Eryilmaz, O. L.; Kim, S. H. Effect of Tribochemistry on Lubricity of DLC Films in Hydrogen. *Surf. Coat. Technol.* 2014, 257, 241–246.
3. Fontaine, J.; Belin, M.; Le Mogne, T.; Grill, A. How to Restore Superlow Friction of DLC: The Healing Effect of Hydrogen Gas. *Tribol. Int.* 2004, 37, 869–877.
4. Srinivasan, S. G.; Duin, A. C. T. Van; Ganesh, P. Development of a ReaxFF Potential for Carbon Condensed Phases and Its Application to the Thermal Fragmentation of a Large Fullerene. *J. Phys. Chem. A* 2015, 119, 571–580.

ELECTROCHEMICAL BEHAVIORS OF LITHIUM INTERCALATION/ DE-INTERCALATION INTO/OUT-OF BORON/CARBON MATERIALS AS ANODES OF LIBS

Masayuki Kawaguchi[1,2*], Seiya Sugisaki[2], Yukako Kouge[2], Kazuma Gotoh[3]

[1] *Fundamental Electronics Research Institute, Osaka Electro-Communication University, Neyagawa, Japan*
[2] *Graduate School of Engineering, Osaka Electro-Communication University, Neyagawa, Japan*
[3] *Department of Chemistry, Okayama University, Okayama, Japan*

* kawaguti@osakac.ac.jp

Introduction

Graphite-like layered materials composed of boron and carbon (B/C materials), and boron, carbon and nitrogen (B/C/N materials) are expected to be new host materials for secondary batteries and new semiconductors. B/C and B/C/N materials showed reversible intercalation and de-intercalation of lithium (Li) and sodium (Na) ions into and out of their layers on their discharge and charge cycles[1,2], which can be applied to anodes of Li and Na ion batteries.

Since Li ion batteries (LIBs) have been used for high energy devices such as electric vehicles, electrodes having higher capacities with better rate performances will be needed. In this study, electrochemical intercalation of Li into B/C materials as anodes of LIBs have been investigated and compared with that of graphite. We have considered an interaction between intercalated Li ions and B atoms in host B/C materials through ^{11}B and ^{7}Li-NMR spectroscopy.

Materials and Methods

B/C materials were prepared through the CVD method using boron trichloride (BCl_3) and ethylene (C_2H_4) as starting materials[2]. B/C materials were mixed with acetylene black (AB) as a conductive additive and polyvinylidene difluoride (PVDF) as a binder with a ratio of B/C material : AB : PVDF = 20 : 2 : 1 in wt%. The electrochemical measurements were performed using three electrode cell. Pieces of Li metals were used as counter and reference electrodes. The electrolytic solution was 1 M-$LiPF_6$ dissolved in mixed solvent of ethylene carbonate (EC) and diethyl carbonate (DEC) with a 1 : 1 volume ratio. ^{11}B and ^{7}Li-NMR measurements were performed for the intercalation compounds prepared through the electrochemical method.

Results and Discussion

B/C materials having the compositions of BC_{22}, BC_{19}, BC_{11} and $BC_{7.9}$ were prepared at 1270 K (or 1170 K) with the molar ratio of staring materials of BCl_3:C_2H_4 = 1:12, 1:6, 1:3 and 4:3, respectively. These materials showed larger reversible capacities than that of graphite (theoretical capacity: 372 mAh/g). Particularly the material with the composition of BC_{19} showed the highest reversible capacities of 540 mAh/g among the materials prepared in this study. Potentials at the beginning of discharge (intercalation of Li^+ ion) in the cases of B/C materials were higher than those of graphite and hard carbon, which could lead to higher reversible capacities. A formation of stage structures was observed during the discharge (electrochemical intercalation) in 1 M-$LiPF_6$/EC+DEC (1:1) electrolyte solution, which was not clearly observed in the case of hard carbon. The first stage compounds prepared through the electrochemical method had larger *d*-spacings (d = 0.39 nm) than that of Li-intercalated graphite (d = 0.37 nm).

Figure 1 shows rate capabilities of B/C materials compared with that of graphite. The material with the composition of BC_{19} showed a charge capacity of about 280 $mAhg^{-1}$ by using a rate of 30 C, which was the highest value among those of B/C materials and commercially available graphite. These electrochemical performances as the anodes of LIBs depended on the composition and the crystallinity. BC_{19} with a moderate crystallite size ($L_{c(002)}$ = 15 nm) showed the best electrochemical performances.

Figure 2 indicates ^{11}B-NMR spectra of the Li-intercalated $BC_{7.9}$ and the original $BC_{7.9}$. The chemical shift of ^{11}B-NMR changed from the original position (+21 ppm) to -4.5 ppm through the intercalation. This result suggests that the intercalated Li supplied electron to the boron in $BC_{7.9}$. This ^{11}B- and ^{7}Li-NMR spectra of Li-intercalated $BC_{7.9}$ suggest that the intercalated Li^+ ion interacted with boron atoms in host $BC_{7.9}$. The interaction of the intercalated Li with boron in B/C materials could make the larger capacities, while the wide *d*-spacing of Li-intercalated B/C materials could make the better rate capabilities as the anodes of LIBs.

Conclusions

B/C materials with a composition of BC_{19} showed the highest reversible capacity about 540 $mAhg^{-1}$ and the highest rate capability among the B/C materials and commercially available graphite.

Acknowledgment

This work was supported by JSPS KAKENHI Grant Number 15H03852.

References

1. Kawaguchi M., Imai Y., Kadowaki N. (2006). Intercalation chemistry of B/C/N materials for anode of Li ion battery. *J. Phys. Chem. Solids* 67, 1084-1090.
2. Yamada K., Ishikawa H., Kamiwaki C., Kawaguchi M. (2015). The role of boron in B/C/N and B/C materials as an anode of sodium ion batteries. *Electrochemistry* 83, 452-458.

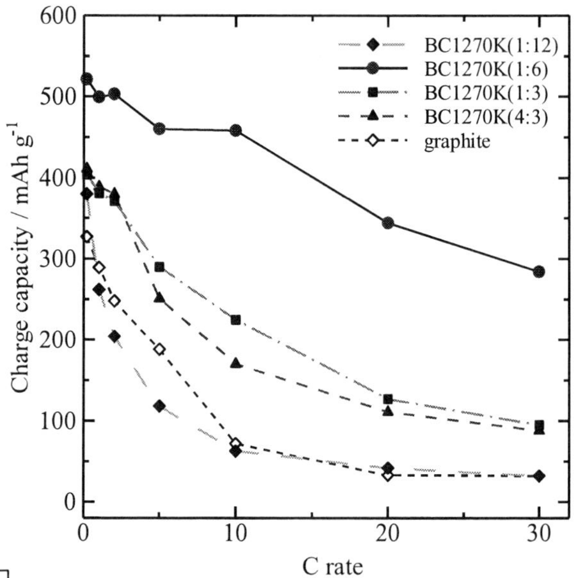

Figure 1. Rate capabilities of B/C materials and commercially available graphite in 1M-$LiPF_6$/ EC+DEC (1:1) electrolyte solution. Composition: BC_{22} for BC1270K(1:12), BC_{19} for BC1270K(1:6), BC_{11} for BC1270K(1:3), and $BC_{7.9}$ for BC1270(4:3).

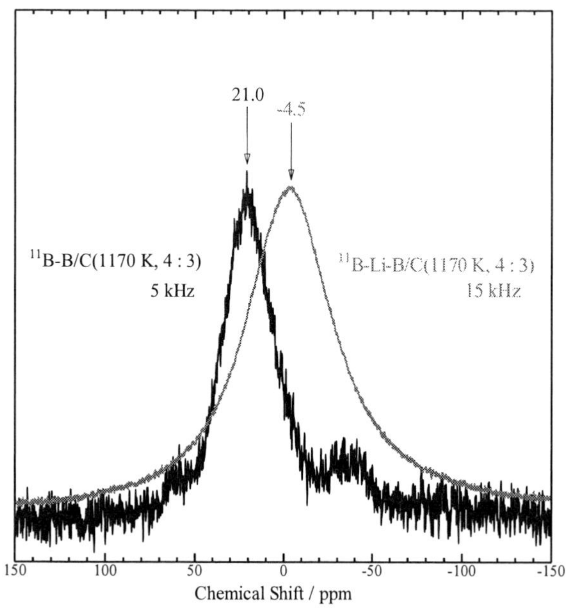

Figure 2. ^{11}B NMR spectra of Li-intercalated B/C and original B/C materials (composition: $BC_{7.9}$).

ELECTROCHEMICAL PERFORMANCE OF N-DOPED SUPERPOROUS ACTIVATED CARBONS IN NON-CONVENTIONAL ELECTROLYTES

María José Mostazo-López[1*], Jakob Krummacher[2], Andrea Balducci[2], Emilia Morallón[3], Diego Cazorla-Amorós[1]

[1]*Instituto Universitario de Materiales, Departamento de Química Inorgánica, Universidad de Alicante, Apartado 99, 03080-Alicante, Spain*
[2]*Institute for Technical Chemistry and Environmental Chemistry and Center for Energy and Environmental Chemistry (CEEC), Philosophenweg 7a, Jena, Germany*
[3]*Instituto Universitario de Materiales, Departamento de Química Física, Universidad de Alicante, Apartado 99, 03080-Alicante, Spain*

* mj.mostazo@ua.es

Introduction

Supercapacitors are energy storage devices that provide high power density, long cycle life and can be used in a high temperature range. However, it is necessary to increase their energy density in order to expand their applications. One of the main strategies for improving their performance is the use of innovative electrolytes, such as those based on ionic liquids or new conducting salts (in organic solvents), since they can expand the operation voltage to values larger than 3V. On the other hand, carbon materials with well-developed microporosity and adequate surface chemistry are highly desirable since they provide high capacitance and electrochemical stability, avoiding detrimental reactions with the electrolyte and increasing the durability of the device. In this sense, it has been proven that nitrogen-doped carbon materials can improve several properties, such as electrical conductivity or electrochemical stability.

In this work, we propose the nitrogen functionalization of a highly microporous activated carbon (S_{BET}>3000m^2/g) by chemical methods at low temperature for their use as electrodes. The effect of the functionalization treatments on the properties of the carbon materials was analyzed. The electrochemical performance of the materials was evaluated by using non-conventional electrolytes.

Materials and Methods

Activated carbon (KUA) was produced by chemical activation of anthracite with KOH as activating agent (4:1 ratio) at 750°C for 2 hours. KUA was functionalized with nitrogen by following two different approaches[1]. The first one consisted in a 3-step protocol: (i) KUA was oxidized with HNO_3 65% (1g/40 mL) at room temperature for 3 hours; (ii) the oxidized carbon was refluxed for 5 hours in a 10% $SOCl_2$/Toluene solution (1g/60mL) and (iii) further treated in a 1M NH_4NO_3/DMF/Pyridine solution (1g/300 mL) (sample named as KUA-$CONH_2$). The second approach consisted in applying the step 3 of the first protocol directly over KUA (sample named as KUA-N).

Porous texture of the samples was characterized by N_2 adsorption/desorption at -196 °C and by CO_2 adsorption at 0° C. XPS analyses were performed in a VG-Microtech Multilab 3000 spectrometer (Al anode). The electrochemical performance of the activated carbons was evaluated by using Swagelok-type cells (2 and 3-electrode configuration) in non-conventional electrolytes (1M 1-butyl-1-methylpyrrolydinium tetrafluoroborate in propylene carbonate (1M $Pyr_{14}BF_4$/PC) and 1M 1-butyl-1-methylpyrrolydinium bis-(tri-fluoromethylsulfonl)imide in propylene carbonate (1M Pyr_{14}TFSI/PC). Electrodes were prepared mixing the sample with acetylene black and PTFE (as binder) in a ratio of 85:10:5 (w/w). The 2-electrode cells were constructed by placing the electrodes (~1mg, 6mm diameter) onto stainless steel collectors and then pressing them against each other using a glass microfiber membrane as separator. In case of 3-electrode cells, the electrode (~5 mg) was pressed against the counter electrode (an activated carbon electrode) and Ag was used as quasi-reference electrode.

Results and Discussion

Table 1 summarizes the physicochemical and electrochemical properties of the samples. XPS analyses show that nitrogen doping was satisfactorily achieved for KUA-N and KUA-$CONH_2$ samples through consumption of oxygen functionalities. Moreover, most of the microporosity of the pristine material remains in the N-doped samples due to the mild conditions of the treatment, allowing the generation of carbon materials with similar porous texture but different functionalities[1].

Table 1. Physicochemical and electrochemical properties of the samples.

Sample	N_{XPS} (at. %)	CO_2 TPD (μmol/g)	CO TPD (μmol/g)	S_{BET} (m^2/g)	V_{DR}^{N2} (cm^3/g)	C_{3E} (F/g)	C_{2E} (F/g)	E (Wh/kg)
KUA	0.3	500	2100	3460	1.19	168	45	59
KUA-N	3.7	700	1610	3450	1.19	163	43	56
KUA-$CONH_2$	4.2	1030	2490	2160	0.74	165	39	50

The electrochemical performance of the samples was tested in 3- and 2- electrode cells in 1M $Pyr_{14}BF_4$/PC and 1M Pyr_{14}TFSI/PC). The materials provide large capacitance (between 160-180 F/g) and a wide potential window in both electrolytes. Table 1 summarizes the main characteristics of the materials in the ionic liquid mixture. Symmetric capacitors were constructed to evaluate their performance at high operation voltage (3.2 V). Fig. 1 illustrates the GCD cycles and the Ragone plot obtained for some of the capacitors. The devices provide high energy values (50-60 Wh/kg, Table 1) in both electrolytes. The effect of nitrogen doping on the durability of the devices was analysed by a cycling test (data not included), evidencing an improvement of the performance of N-doped activated carbons due to the beneficial effect of nitrogen doping at mild conditions.

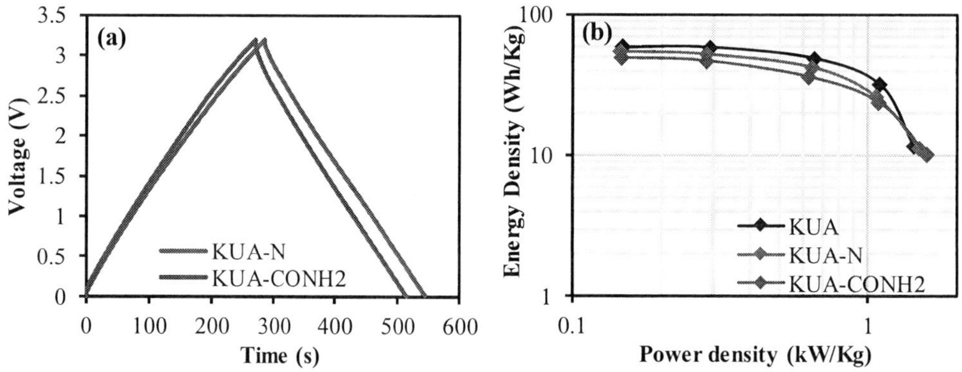

Figure 1. (a) GCD cycles at 0.5 A/g and (b) Ragone plot obtained for the activated carbons-based capacitors. 1M Pyr_{14}TFSI/PC. V = 3.2 V.

Conclusions

Functionalization of an activated carbon with well-developed microporosity by chemical post-treatments has been carried out to introduce nitrogen functionalities while preserving most of the porous texture. The combination of materials with high capacitance and electrolytes with large operation voltage allows obtaining devices with high energy density as well as remarkable durability due to the effect of nitrogen functional groups.

Acknowledgment

The authors thank MINECO and FEDER (CTQ2015-66080-R (MINECO/FEDER), MAT2016-76595-R) and Generalitat Valenciana (ACIF/2015/374, BEFPI/2017/036).

References

1. Mostazo-López M.J., Ruiz-Rosas R., Morallón E., Cazorla-Amorós D. (2016). Nitrogen doped superporous carbon prepared by a mild method. Enhancement of supercapacitor performance. *Int. J. Hydrogen Energy* 41, I9691-I9701.

Just add water and table salt: new perspectives on the electrochemical exfoliation of high quality graphene

Jose M. Munuera[1]*
Juan I. Paredes[1], Marina Enterría[2], Ana Pagán[3], Silvia Villar-Rodil[1], Manuel F. R. Pereira, Jose I. Martins[2,4], José L. Figueiredo[2], José L. Cenis[3], Amelia Martínez-Alonso[1], Juan M. D. Tascón[1]

[1] *Instituto Nacional del Carbón, INCAR-CSIC, Oviedo, Spain*
[2] *Faculdade de Engenharia, Universidade do Porto, Porto, Portugal*
[3] *Instituto Murciano de Investigación y Desarrollo Agrario y Alimentario (IMIDA), La Alberca, Spain*
[4] *LAB2PT- Laboratório de Paisagens, Património e Território, Universidade do Minho, Braga, Portugal*

* j.munuera@incar.csic.es

Introduction

Graphene materials have shown great promise in many technological applications, but their large-scale production and processing by simple and cost-effective means still consitute a hurdle in the path of their practical implementation. Electrochemical exfoliation has emerged as a practical and scalable method to this end. It relies on the delamination of graphite electrodes in proper electrolytes, where the ionic intercalation leads to the expansion and exfoliation of graphite to give graphene nanosheets. Anodic exfoliation in particular(where graphite is set as the anode) is a promising route, since it can be carried out in water and has a better yield and exfoliation degree than cathodic delamination. However, this method tends to yield substantially oxidized nanosheets due to the anodic oxidation of water molecules. Oxidation levels typically lie in the 0.10-0.20 range (O/C atomic ratio), but several strategies have been developed to avert this unwanted effect. Here, we discuss several strategies to reduce the oxygen content of anodically exfoliated graphene. Such strategies rely on a judicious choice of graphite material, the use of proper electrolytes or electrolyte additives as well as the electrolyte concentration.

Materials and Methods

Graphite powder (Sigma Aldrich), flakes (Mersen), foil (Mersen) and highly oriented pyrolytic graphite (HOPG, Advanced Ceramics) were used as anode in the anodic exfoliation processes. Platinum wire and foil were employed as the cathode, and a DC power supply (Agilent 6614C) was used to set the bias voltage (10 V). Exfoliation was carried out in aqueous electrolyte for 3 to 60 minutes. The resulting expanded graphite product was collected, filtered, washed with Milli-Q water and dried at room temperature under reduced pressure. The materials were used as dried powders or dispersed via sonication in the liquid phase, either in N,N-dimethylformamide (DMF), water-isopropyl alcohol mixtures or aqueous surfactant solutions.

Results and Discussion

As a first step in the investigation, we examined the effect of graphite type used for anodic exfoliation. Graphite powder, flakes, foil and HOPG were exfoliated in aqueous sulfate solutions. Graphene nanosheets obtained thereof were seen to differ in their oxygen content and exfoliation degree. Small-particle graphites (powder and flakes) yielded poorly exfoliated graphene, while graphite foil and HOPG afforded thinner, but more oxidized nanosheets. The oxidation degree was higher for HOPG, because its dense structure required a more exntensive oxidation to facilitate a proper ion intercalation. On the other hand, the wrinkled, imperfectly stacked morphology of the foil made the interaction and (and hence the delamination) process more effcicient without requiring extensive oxidation.

Two strategies were explored to decrease the O/C ratio of anodic graphene. The first was based on the use of oxidizable electrolytes in place of sulfate anions, and the second relied on the use of antioxidants as additives to the sulfate electrolyte. In both cases, the central idea is to include chemical species in the electrolytic medium that act as sacrificial agents, so that they become oxidized during exfoliation instead of the graphene nanosheets. To this end, we tested aromatic sulfonates as electrolytes, and the results indicated that some of them got indeed oxidized and avoided oxidation the nanosheets (O/C ratio as low as ~0.02). . Sodium halides were used for the same purpose. Although they were previously reported as unsuccessful for anodic exfoliation of graphite, we observed that a proper choice of both anode material (graphite foil) and sodium halide concentration, graphenes with relatively low O/C ratio (0.06) could be obtained. Finally, chloride anions (in the form of NaCl or KCl) were used as an additive of sulfate-based electrolytes for the anodic exfoliation process. When used in optimal quantities, they were seen to function as particularly efficient additives that prevented graphene oxidation without negatively interfering with the ability of the sulfate anion to induce delamination, affording graphene nanosheets with an O/C ratio down to 0.02. This finding can have important practical applications in the large-scale production of high quality graphene nanosheets, as only simple, cheap and abundant chemical species (sulfate salts and NaCl) are required in their preparation.

Table 1. O/C ratios for graphenes obtained through anodic exfoliation in different aqueous electrolytes

Electrolyte	O/C ratio
Potassium sulfate	0.11
Sodium 1,5-naphthalenedisulfonate	0.02
Sodium chloride	0.06
Sodium sulfate + sodium chloride	0.02

Figure 1. As-exfoliated graphene dry powder. SEM image of the same material (inset).

Conclusions

We have demonstrated a number of successful strategies for the preparation of anodically exfoliated graphene that avoid the (generally unwanted) oxidation of the nanosheets typically associated to this delamination approach. Such strategies relied on the use of proper electrolytes and/or electrolyte additives that act as sacrificial agents in the anodic oxidation processes.

Acknowledgment

We acknowledge funding through grant MAT2015-69844-R from the Spanish MINECO and the ERDF and pre-doctoral contract FPU14/00792 from the Spanish MECD.

References

1. J.M. Munuera, J.I. Paredes, S. Villar-Rodil, M. Ayán-Varela, A. Pagán, S.D. Aznar-Cervantes, J.L. Cenis, A. Martínez-Alonso, J.M.D. Tascón (2015) High quality, low oxygen content and biocompatible graphene nanosheets obtained by anodic exfoliation of different graphite types. *Carbon* 94, 729-739.
2. J. M. Munuera, J. I. Paredes, S. Villar-Rodil, M. Ayán-Varela, A. Martínez-Alonso, J. M. D. Tascón (2016) Electrolytic exfoliation of graphite in water with multifunctional electrolytes: en route towards high quality, oxide-free graphene flakes. *Nanoscale* 8, 2982-2998.
3. J. M. Munuera J. I. Paredes, M. Enterría, A. Pagán, S. Villar-Rodil, M. F. R. Pereira, J. I. Martins, J. L. Figueiredo, J. L. Cenis, A. Martínez-Alonso, J. M. D. Tascón (2017). Electrochemical exfoliation of graphite in aqueous sodium halide electrolytes toward low oxygen content graphene for energy and environmental applications. *ACS Applied Materials and Interfaces* 9, 24085-24099.
4. J.M. Munuera, J.I. Paredes, S. Villar-Rodil, A. Castro-Muñiz, A. Martínez-Alonso, J.M.D. Tascón (2018) High quality, low-oxidized graphene via anodic exfoliation with table salt as an efficient oxidation-preventing co-electrolyte for water/oil remediation and capacitive energy storage applications. *Applied Materials Today* 11, 246–254.

FLOW PATTERN ANAYSIS OF THE HORIZONTAL QUARTZ TUBE REACTOR FOR CNT FIBER SYNTHESIS

Kun-Hong Lee*, Taeseon Lee, Taewoong Um, Sung-Hyun Lee, In Seok Kang

Department of Chemical Engineering. Pohang University of Science and Technology (POSTECH), Pohang, South Korea

*Presenting author's e-mail: ce20047@postech.ac.kr

Introduction

Since from the injection of raw materials to the continuous synthesis of CNT fibers occurs in a single reactor with a typical residence time of less than a minute, direct spinning has an advantage in the scale-up for commercial production. In reality, one has to consider thermodynamics, fluid mechanics, and heat transfer for the scale up of a reactor. Especially, analysis of flow pattern inside a reactor is very important because the residence time distribution (RTD) is directly influenced by the flow pattern. RTD affects on the growth and the lifetime of catalyst particles, and consequently the quality of synthesized CNT fibers.

Quartz tube reactor has been used for the direct spinning of CNT fibers. The overall flow rate of gases in direct spinning is generally low, and typical Reynolds number is in the order of 10s~100s. Therefore, most of previous researchers assumed that the flow inside the tube reactor was a laminar flow. However, existence of flow circulations inside the tube reactor was predicted in the recent publication based on the computational fluid dynamics (CFD) simulations. Although the simulation revealed important circulation phenomenon, experimental verification was incomplete.

This paper is complimentary of the excellent work of Guangfeng Hou et al. in a sense that flow circulations at the inlet and the outlet of a horizontal quartz tube reactor are experimentally verified using flow visualization technique.

Materials and Methods

A horizontal quartz tube furnace was used for the flow visualization experiment. The maximum temperature inside the quartz tube reactor was 1175°C, and high purity Argon was used at a flow rate of 1000 and 4000 sccm. The coffee beans inside the tube reactor generated smoke when heated. Circulation of the flow was recorded by the four video cameras positioned as in the following Figure.

Results and Discussion

There were longitudinal (horizontal) and cross - sectional (vertical) temperature distributions inside the reactor. There was a large temperature difference across the cross section at the furnace inlet (80°C) or the outlet (60°C), while there was no temperature difference across the cross section at the middle of the reactor.

	-15	-10	-5	0	5	10	15	20	25	30	35	40	45	50	55	60	65	70	75	80	85	90	95	100	105	110	115	120	125
Top	430	557	685	805	940	1085	1154	1170	1173	1175	1170	1143	1025	883	734	647	523	432	383	338	296	249	199	168	144	127	115	105	110
Side	307	465	646	772	916	1074	1153	1168	1173	1174	1166	1133	1009	834	690	542	369	272	231	202	161	138	111	95	75	72	60	44	29
center	340	489	633	753	906	1160	1141	1162	1168	1167	1160	1128	1009	848	704	549	408	297	243	202	185	161	138	120	101	85	82	51	29
Bottom	266	435	606	724	892	1056	1144	1168	1174	1173	1163	1126	988	817	680	527	331	237	163	132	110	103	71	56	51	48	42	31	27

Two flow circulations were observed at the furnace inlet and outlet, and the in-flow of external air from the tube outlet was confirmed in the flow visualization experiment. These findings were schematically shown in the following Figure, and short movies will be presented at the conference presentation.

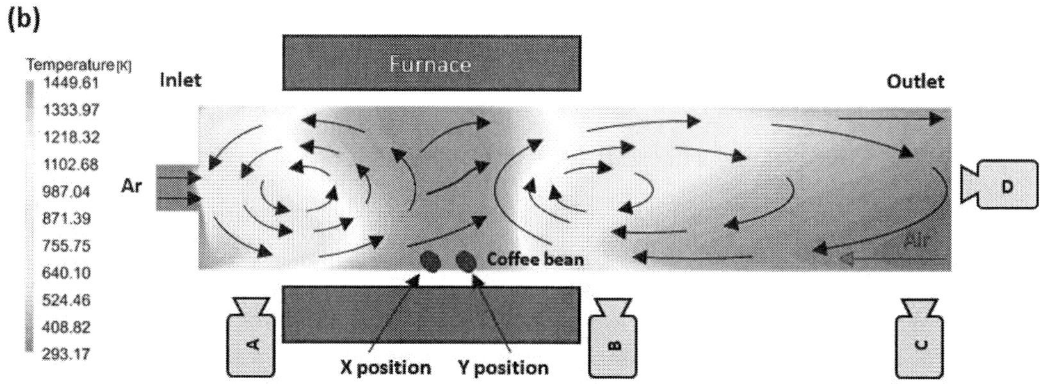

It was also observed that the spatial distribution of catalyst particle size became complex. We think this is a consequence of the complex flow pattern inside the horizontal quartz tube reactor. It was indirectly proved by our own simulations that the main driving force of these circulations was the buoyancy force due to the temperature difference in the vertical direction as well as the axial direction of the horizontal quartz tube reactor.

Conclusions

In a horizontal quartz tube reactor, flow is not a simple laminar flow even with a small Reynolds Number when there is a significant temperature distribution inside the reactor. Instead, complicated flow pattern is caused by buoyancy force due to the temperature difference in the vertical direction as well as in the axial direction of the reactor. Important characteristics of these complex flow pattern can be accurately predictable using CFD calculations. (We demonstrated it experimentally in this work.) The effect of buoyancy force would become greater with the increase of the size of the reactor, so that it should be considered seriously for the scale-up of the reactor.

Acknowledgment

This work was funded by a grant from the Korea Institute of Science and Technology Open Research Program (grant no. 2E27820).

References

1. Hou, G., Su, R., Wang, A., Ng, V., Li, W., Song, Y., Zhang, L., Sundaram, M., Shanov, V., Mast, D. Lashmore, D., Schulz, M., Liu, Y. (2016). The effect of a convection vortex on sock formation in the floating catalyst method for carbon nanotube synthesis. Carbon. 102, 513-9

ELUCIDATION OF RELATIONSHIPS BETWEEN CARBON STRUCTURE AND ANION INTERCALATION IN DUAL-ION BATTERIES

Andreas Heckmann[1*], Olga Fromm[1], Uta Rodehorst[1], Martin Winter[1,2], Tobias Placke[1]

[1] *University of Münster, MEET Battery Research Center, Corrensstraße 46, 48149 Münster, Germany*
[2] *Helmholtz-Institute Münster, IEK-12, Forschungszentrum Jülich GmbH, Corrensstraße 46, 48149 Münster, Germany*

*Presenting author's e-mail: andreas.heckmann@uni-muenster.de

Introduction

The research activities of anion intercalation into carbonaceous samples gained a renaissance due to the rising interest in dual-ion cell chemistry, which obtained a considerable advance during the last 5 years, especially in respect to the cycling performance and the possibility to use graphitic carbon-based electrodes for both, the anode and the cathode.[1, 2] Therefore, the dual-ion battery can be considered as a promising and green alternative to lithium ion batteries for stationary energy storage systems. Here, we present comprehensive insights into structure – property relationships of carbon electrodes possessing a different degree of graphitization and their anion uptake capabilities. A particular focus is set on the preparation and characterization of graphitic carbon samples at different heat treatment temperatures and their electrochemical performance as cathode material in dual-ion batteries.

Materials and Methods

Two non-graphitic and two graphitic carbons were heat treated under argon atmosphere in the temperature interval of 2200-2800 °C.

Material parameters, such as morphology, crystallinity, particle size, BET- and DFT-surface areas, were determined by scanning electron microscopy (SEM), X-ray powder diffraction (XRD) and nitrogen adsorption measurements.

In order to correlate obtained material parameters with the anion intercalation properties, constant current cycling (CCC), current rate (C-Rate) and constant current/constant potential (CCCP) charge/discharge measurements were performed in lab-scale three-electrode Swagelok type T-cells.

Results and Discussion

The carbons, heat treated between 2200 °C and 2800 °C, demonstrate graphitization degrees from ≈30 % to ≈90 % (**Figure 1**).

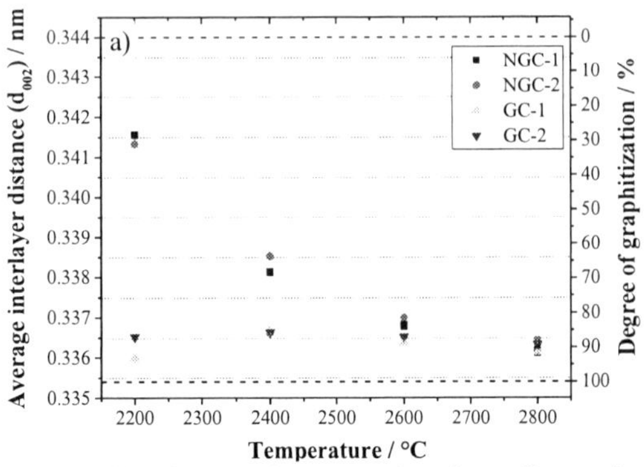

Figure 1: Development of the degree of graphitization and the average interlayer distance in dependence of the heat treatment temperature for two non-graphitic carbons (NGC-1, NGC-2) and two graphitic carbons (GC-1, GC-2).[3]

Electrochemical constant current cycling experiments reveal a linear relation of the degree of graphitization and the specific discharge capacity. For increased cycling temperatures (60 °C), specific discharge capacities of ≈75 mAh g^{-1} and ≈100 mAh g^{-1} are found for carbons providing a graphitization degree of ≈30 % and ≈90 %, respectively (**Figure 2**).

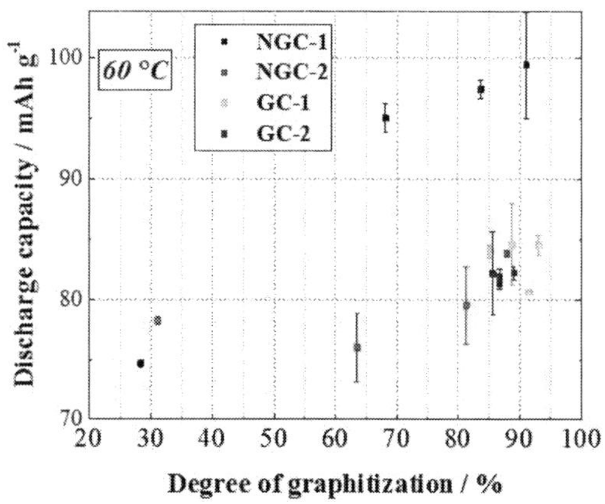

Figure 2: Development of the specific capacities during de-intercalation of TFSI anions in dependence of the heat treatment temperature at a specific charge/discharge current of 50 mA g^{-1} and an operating temperature of 60 °C. (Potential range: 3.4 V – 5.0 V *vs.* Li/Li^{+}).[3]

Further experiments demonstrate that the staging mechanism the pre-dominant storage mechanism for all investigated graphitic carbon electrodes.

Conclusions

The systematic study on the preparation and characterization of graphitic carbon materials revealed a strong dependency of the degree of graphitization (DoG) with the specific discharge (= de-intercalation) capacity. Samples possessing a higher DoG showed a higher specific discharge capacity, which leads to the assumption that a fully graphitized carbon (≈ 100 % DoG) yield the highest value of anion storage.

Acknowledgment

The authors wish to thank the German Federal Ministry of Education and Research (BMBF) for funding this work in the project "Dual-Carb" (03XP0118). Furthermore, we thank H.C. Carbon for providing the carbonaceous sample "MECHANO-REM 5".

References

[1] S. Rothermel, P. Meister, G. Schmuelling, O. Fromm, H.W. Meyer, S. Nowak, M. Winter, T. Placke, Dual-Graphite Cells Based on the Reversible Intercalation of Bis(trifluoromethanesulfonyl)imide Anions from an Ionic Liquid Electrolyte, Energy & Environmental Science 7(10) (2014) 3412-3423.

[2] A. Heckmann, J. Thienenkamp, K. Beltrop, M. Winter, G. Brunklaus, T. Placke, Towards High-Performance Dual-Graphite Batteries using Highly Concentrated Organic Electrolytes, Electrochimica Acta 260 (2018) 514-525.

[3] A. Heckmann, O. Fromm, U. Rodehorst, P. Münster, M. Winter, T. Placke, New Insights into Electrochemical Anion Intercalation into Carbonaceous Materials for Dual-Ion Batteries: Impact of the Graphitization Degree, Carbon 131 (2018) 201-212.

DEVELOPMENT OF NOVEL PAN-BASED ACTIVATED CARBON FIBERS

Masaaki Yoshikawa[1*], Junichi Hayashi[2]

[1] *Energy Technology Laboratories, Osaka Gas co., ltd, Osaka, Japan*
[2]*Kansai University, Osaka, Japan*

yoshikaw@osakagas.co.jp

Introduction

Global production of Poly acrylonitrile (PAN) based carbon fiber has reached to 70,000t/y in 2017. During this mass production, we can access to various grade of PAN carbon fiber from several companies, and we found some of them can be activated into activated carbon fibers (ACF). In our previous work, it was observed that the addition of nitrogen compounds to carbon raw material formed high specific surface area carbons by K_2CO_3 activation. Accordingly, we investigated both steam activation and chemical activation by K_2CO_3 of PAN fibers.

Materials and Methods

We tested various grade of PAN carbon fiber from several companies, and chose one kind of them for this work which has stable quality and reasonable price. We did both steam activation and chemical activation at high temperature over 900 degree C in the electric furnace. In the case of chemical activation, K_2CO_3 was dissolver in water and impregnated in the PAN carbon fiber. The weight ratio of K_2CO_3/Carbon fiber was set at 0.5 ~2.0. After chemical activation, the activated material was washed by water until neutral and dried till few water content. The porous properties of the products were evaluated through N_2 adsorption method at 77K by Micromeritics ASAP2420.

Results and Discussion

The steam activation of PAN carbon fiber at 900~1000 degree C resulted in making ACF which has specific surface area of 400~800m^2/g, agreed with our previous. We observed adsorption rate of NO gas over PAN based ACF 2 times faster than the conventional pitch based ACF. This adsorption rate is considered to be the influence of both pore size and nitrogen functional group from PAN. The chemical activation of PAN carbon fiber is summarized in **Table 1.** We found the formation of ACF of surface area over 3,000m^2/g at the K_2CO_3/Carbon=2.0. STEM image of steam and chemical activated ACF were shown in **Figure 1.** The surface structure of these ACF were different obviously, and the chemical activation made high porosity inside the carbon fiber, which is considered as intercalation of K compounds into graphitic layers.

Table 1. Properties of Chemical Activated ACF

K_2CO_3/Carbon fiber (wt)	0.5	1.0	2.0
Specific surface area (m^2/g)	455	1025	3175
Meso pore ratio (%)	11.4	32.4	76.8

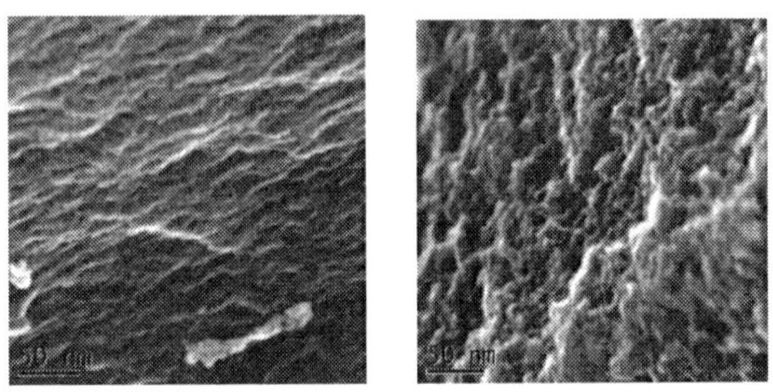

Steam activated **Chemical activated**

Figure 1. STEM image of PAN-ACF

Conclusions

The high porosity and high surface area of chemical activated ACF is seemed to be the effect of nitrogen included in the raw material PAN fiber.

3D PRINTED GUIDELINES FROM NANOFIBRILLATED CELLULOSE/CARBON NANOTUBES INKS FOR APPLICATION IN NEURAL TISSUE ENGINEERING

Volodymyr Kuzmenko[1,2*], Erdem Karabulut[2,3], Elin Pernevik[3], Peter Enoksson[1,2] and Paul Gatenholm[2,3]

[1]*Department of Microtechnology and Nanoscience, Chalmers University of Technology, Gothenburg, Sweden*
[2]*Wallenberg Wood Science Center, Chalmers University of Technology, Gothenburg, Sweden*
[3]*Department of Chemistry and Chemical Engineering, Chalmers University of Technology, Gothenburg, Sweden*

*Presenting author's e-mail: kuzmenko@chalmers.se

Introduction

Currently, neurodegenerative disorders and spinal cord injuries stand for an expanding public health problem that costed almost $818 billion in 2015.[1] Neural tissue engineering (TE) is a method that can be utilized to grow healthy patient-specific neural tissues to replace defective ones, and, hopefully, to completely solve the problem mentioned above. Recently, nanocellulose-based materials emerged as biocompatible and sustainable candidates for low-cost TE scaffolds that can be shaped for specific patient needs via 3D printing technique.[2,3] Moreover, charged nanofibrillated cellulose (NFC) can homogeneously disperse carbon nanotubes (CNTs) in aqueous media by forming electrostatically stable NFC/CNT structures,[4] which results in electrically conductive inks suitable for 3D printing of scaffolds that transfer electrochemical signals between neural cells.

In this work, we synthesize conductive NFC/CNT inks, use them for 3D printing of conductive guidelines and demonstrate how these guidelines stimulate attachment, proliferation, growth and differentiation of human-derived neuroblastoma cells *in vitro*.

Materials and Methods

2.4 wt% aqueous dispersion of carboxymethylated NFC (charge density of 515 μeq g^{-1}) was mixed with 1 wt% aqueous SWCNT dispersion (1.0–3.0 at% COO^-H^+) and homogenized with ULTRA-TURRAX mixer at 8000 rpm for 10 min. To get proper miscibility between components, the mixture's pH was tuned to 8.5. Centrifugation and air-drying were used to regulate the final concentration of an ink. The guidelines were created using a 3D printer (RegenHU 3D Discovery) with a piezoelectric microvalve and a 300 μm nozzle. The dispensing pressure was set to 65 kPa, the dosing drop distance to 0.05 mm, and the printing speed to 10 mm s^{-1}.

The SH-SHY5Y human neuroblastoma cells were expanded until passage 22 in growth medium, maintained in a humidified atmosphere at 37 °C and 95% air/5% CO_2. The differentiation medium with a serum concentration of 1 wt% was also supplemented with 10 μM retinoic acid.

Results and Discussion

To make a printable conductive ink, CNTs were homogeneously dispersed inside NFC hydrogel to form a viscous ink that can be 3D printed into guidelines with a diameter below 1 mm and electrical conductivity of $3.8 \cdot 10^{-1}$ S cm^{-1} after ink solidification (**Fig. 1A**). Sufficient fraction of CNTs makes it possible to reach a percolation threshold in the printed structures.

Afterwards, cellulose-based materials with conductive guidelines were tested as scaffolds for neural TE (**Fig. 1B**). According to our study, these guidelines show clear ability to enhance and guide neural cell attachment, proliferation, growth, network formation and differentiation at least within 27 days of cell culturing. Along with an appropriate nanotopography and surface chemistry, electrical conductivity of the composite guidelines apparently have the greatest influence on the behaviour of neural cells.[5]

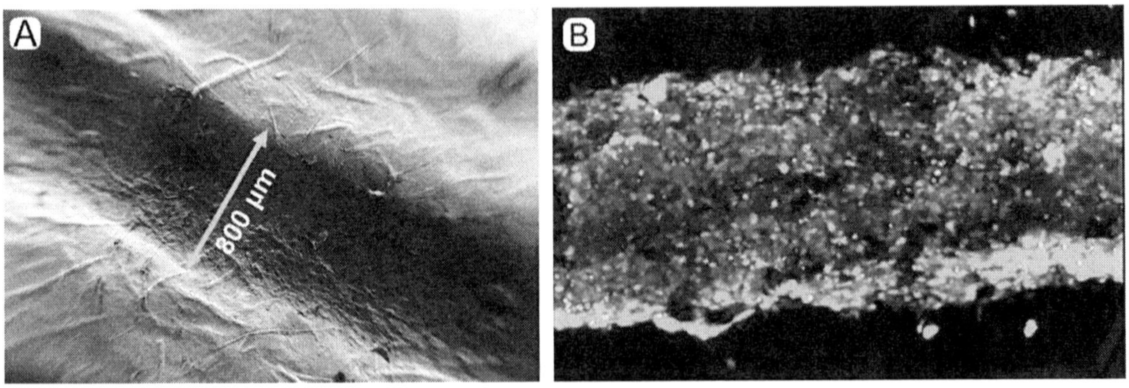

Figure 1. A) SEM image of a typical guideline 3D printed on the NFC substrate using the conductive NFC/CNT ink (mag=100×). B) Live/Dead staining of SH-SY5Y human neuroblastoma cells after 10 days of culturing on the printed conductive guidelines. Green fluorescent cells and red fluorescent cells indicate alive and dead cells, respectively.

Conclusions

Composite NFC/CNT inks were optimized for 3D printing by tuning the surface charges of NFC and CNTs with pH regulation. The resulting 3D printed conductive NFC/CNT guidelines were utilized for neural TE. The guidelines showed extraordinary capability to enhance cell attachment and proliferation.

Acknowledgment

The Wallenberg Wood Science Center funded by Knut and Alice Wallenberg Foundation is greatly acknowledged for their financial support. We also thank Innventia that kindly supplied the CNFs. We appreciate the help of Anne Wendel, Anders Mårtensson, Kajsa Markstedt, Daniel Hägg and Sara Johannesson with the XPS, AFM, rheology and cell analyses.

References

1. Prince M., Wimo A., Guerchet M., Ali G.-C., Wu Y.-T. and Prina M. (2015). *World Alzheimer report 2015. The global impact of dementia*. London: Alzheimer's Disease International (ADI).
2. Markstedt K., Mantas A., Tournier I., Martínez Ávila H., Hägg D. and Gatenholm, P. (2015). 3D bioprinting human chondrocytes with nanocellulose-alginate bioink for cartilage tissue engineering applications. *Biomacromolecules* 16, 1489-1496.
3. Håkansson K.M.O., Henriksson I.C., de la Peña Vázquez C., Kuzmenko V., Markstedt K., Enoksson, P. and Gatenholm P. (2016). Solidification of 3D printed nanofibril hydrogels into functional 3D cellulose structures. *Advanced Materials Technologies* 1, 1600096.
4. Hamedi M.M., Hajian A., Fall A.B., Håkansson K., Salajkova M., Lundell F., Wågberg L. and Berglund L.A. (2014). Highly conducting, strong nanocomposites based on nanocellulose-assisted aqueous dispersions of single-wall carbon nanotubes. *ACS Nano* 8, 2467-2476.
5. Kuzmenko V., Karabulut E., Pernevik E., Enoksson P. and Gatenholm P. (2018). Tailor-made conductive inks from cellulose nanofibrils for 3D printing of neural guidelines. *Carbohydrate Polymers* 189, 22-30.

CARBIDE-DERIVED CARBON BEADS WITH TUNABLE NANOPORES FROM CONTINUOUSLY PRODUCED POLYSILSESQUIOXANES FOR ENERGY STORAGE DEVICES

Benjamin Krüner[1,2*], Christina Odenwald[3], Guido Kickelbick[3], Volker Presser[1,2]

[1]*INM – Leibniz Institute for New Materials, Saarbrücken, Germany*
[2]*Department for Materials Science and Engineering, Saarland University, Saarbrücken, Germany*
[3]*Inorganic Solid State Chemistry, Saarland University, Saarbrücken, Germany*

*Presenting author's e-mail: benjamin.kruener@leibniz-inm.de

Introduction

The MicroJet reactor technique is an excellent continuous method to produce spherical and homogeneous organically modified silica particles on a large scale (10-15 g/min). We applied this method to manufacture polyorganosilsesquioxanes with different ratios of functional groups as a carbon source. The polymer beads were pyrolyzed to obtain silicon oxycarbide. These polymer-derived ceramics (PDC) can be used as the anode for lithium-ion batteries and are highly suited as a precursor for carbide-derived carbon (CDC) synthesis. Chlorine etching of PDC at high temperatures removed silicon and oxygen, yielding the formation of nanoporous carbon. Pure poly(phenyl-silsesquioxane) spheres lost their shape during the thermal process because of further condensation reactions. Yet, the spherical shape was conserved during thermal processing after adding vinyl functionalities, without an additional crosslinking agent.

Materials and Methods

Vinyltrimethoxysilane and phenyltrimethoxysilane were used to produce the polyorgano-silsesquioxanes beads, by an ammonia catalyzed a sol-gel reaction with the MircoJet reactor technique. We further use a nitrogen-containing (3-anilinopropyl)trimethoxysilane to introduce nitrogen in the structure, using the same method. The materials were fully analyzed amongst others with energy dispersive X-ray photoelectron spectroscopy, Raman spectroscopy, X-ray diffraction, and electron microscopy.

The electrochemical supercapacitor performance was tested in a symmetrical full-cell using 1 M tetraethylammonium tetrafluoroborate in acetonitrile (TEA-BF_4 in ACN) as the electrolyte. The nitrogen-doped carbide-derived carbon materials were tested as well in a half-cell configuration with 1-ethyl-3-methylimidazolium tetrafluoroborate (EMIM-BF_4).

The polymer-derived ceramics were additionally applied as an anode material for lithium-ion batteries in a half-cell vs. lithium using 1 M lithium hexafluorophosphate dissolved in ethylene carbonate/dimethyl carbonate ($LiPF_6$ in EC/DMC).

Results and Discussion

In a first study, we varied the ratio of vinyl and phenyl functionalities, whereby it was possible to control the pore structure and the total CDC yield, enabling an increase from 2 mass% to 22 mass%.[1] The total pore volume was 1.3-2.1 cm^3/g, and the specific surface area ranged at 2014-2114 m^2/g. The high surface area and large pore volume make these materials attractive for high power supercapacitor electrodes. The specific capacitance in 1 M TEA-BF_4 in ACN of the best resulting CDC was 116 F/g (specific energy 25 Wh/kg) at low rate (at 5 mA/g) and still 80 F/g (specific energy of 12 Wh/kg) at ultra-high rate (100 A/g; specific power of 41 kW/kg).

In a second study, we introduced nitrogen in the carbide-derived carbon by using a nitrogen-containing precursor which led to a nitrogen content from 1-7 mass% depending on the chlorination temperature (600-1000 °C). We also synthesized carbide-derived carbon beads without nitrogen using the same conditions. The highest porosity was obtained at a synthesis temperature of 800 °C with a specific surface area of ~2000 m^2/g and a pore volume of ~1.7 cm^3/g. The nitrogen groups increased the specific capacitance in TEA-BF_4 in ACN measured in a half cell to around 150 F/g. The benefits of the nitrogen groups are striking when an ionic liquid (EMIM-BF_4) is used as the electrolyte. Thereby the specific capacitance at -1.5 V vs. carbon increased from around 180 F/g to 215 F/g.

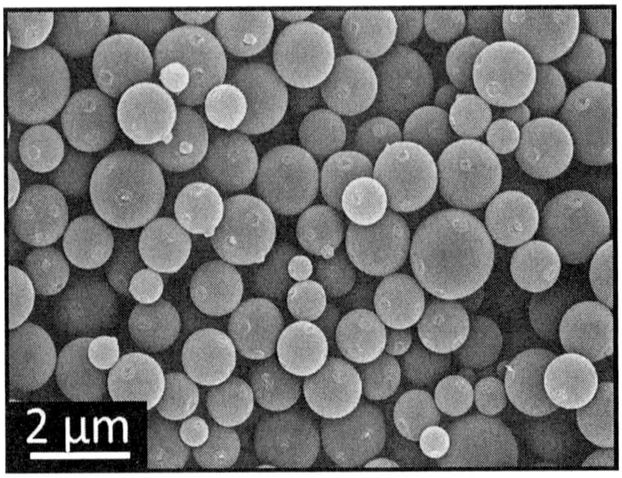

Figure 1. SEM image of the carbide-derived carbon beads.[1]

A third study focused on the use of the silicon oxycarbide, which were obtained by the pyrolysis of the poly-silsequioxanes, as anode material for lithium-ion batteries. Different silanes were used to produce the silicon oxycarbides with different compositions. Thereby, the amount of silicon/carbon bonds, as well as the free carbon content, were modified. A mixture of vinyl and phenyltrimethoxysilane was identified as the best anode material with a stable performance due to the increased carbon content. The 1st delithiation capacity of the most stable material was 922 mAh/g and the capacity retention after 100 cycles was 83% (767 mAh/g).

Conclusions

The continuous MicroJet reactor technique is a facile way to produce spherical polymer particles with a constant quality in large amounts. The polymer beads were found as highly suited for pyrolysis, to obtain SiOC, and chlorine gas treatment, to obtain CDC by removal of non-carbon elements and still conserving the spherical morphology. The porosity of the CDCs can be tuned by the chemical composition of the precursor and the synthesis temperature. In conclusion, the high porosity of the SiOC-CDCs is attractive for electrochemical energy storage with supercapacitors. The excellent rate handling behaviour and the high stability in organic electrolyte documents suitability for common supercapacitors.

References

1. Krüner, B., Odenwald, C., Tolosa, A., Schreiber, A., Aslan, M., Kickelbick, G., Presser, V. (2017). Carbide-derived carbon beads with tunable nanopores from continuously produced polysilsesquioxanes for supercapacitor electrodes. *Sustainable Energy & Fuels* 1, 1588-1600

NANOPOROUS CARBONS FOR SOLAR FUELS PRODUCTION

Juan Matos[1*], Jose Ocares[1,2], Po S. Poon[1], Ximena García[2], María M. Titirici[3], Conchi O. Ania[4]

[1] *Hybrid and Carbon Materials Group, UDT, University of Concepción, Chile.*
[2]*Department of Chemical Engineering, University of Concepción, Concepción, Chile.*
[3] *Queen Mary University of London, School of Engineering and Materials Science, London, UK.*
[4] *POR2E Group, CEMHTI CNRS (UPR3079), Univ. Orleans, Orleans, France.*

*Presenting author's e-mail: jmatoslale@gmail.com

Introduction

Solar fuels are a promising strategy since they are produced from simple and abundant molecules using a renewable energy source such as sunlight. However, the efficiency of photocatalysts is still low and far for the practical application. The H_2 production by direct water splitting[1] using TiO_2 and the CO_2 reduction[2] are the two main reactions concerning with solar fuels. It has been widely showed that noble metal/graphene-based catalysts are suitable for these reactions[1,2]. A third way poorly studied correspond to the photo assisted valorization of molecules such as furfuryl alcohol (FA) and 5-hydroximethyl furfural (5-HMF) because they are important products of the biorefinery industry[3]. In addition, glycerol (Gly) is the main by-product of biodiesel industry[4] with an important economic devaluation[5]. The FA and Gly photo assisted valorisation to selectively produce aldehydes or ketones is an important strategy to obtain more valuable compounds such as five or six-member dioxane-based carbocycles. In this sense, this work explores a photocatalytic approach for the valorization of FA, 5-HMF and Gly by using C-doped and Cu-containing TiO_2-based catalysts, under solar light.

Materials and Methods

C-doped TiO_2 photocatalyst was prepared by solvothermal synthesis by mixing furfural and titanium isopropoxide in ethanol. The solution was sealed inside an autoclave and heated at 180 ºC for 16 h. The solid was filtered and washed in absolute ethanol and dried under static-air at 100 ºC for 2 h. The material was denoted as TiO_2-C. Cu-containing photocatalyst was prepared by adding $Cu(acac)_2$ to the mixture of furfural and Ti-alkoxide (5 wt.% Cu) and proceeding as above. The sample was denoted as Cu@TiO_2-C. The catalysts were characterized by XRD, TEM, gas adsorption, and surface pH. The photocatalytic tests were performed using a Pyrex open-to-air photoreactor with 125 mL of FA or Gly solution and 125 mg of catalyst. Irradiation was provided by a solar simulator (250W Xe-lamp). FA and 5-HMF solutions were used from the remnant liquid from the solvothermal synthesis. For the Gly photocatalytic tests, a solution of 20% v/v was prepared in distilled water. UV-vis spectroscopy was used to follow the reaction in both target molecules and for the case of Gly, the appearance of the product was followed by changes in the transmittance peaks observed in FTIR.

Results and Discussion

Table 1 displays some properties of the catalysts. The amount of Cu measured was ca. 4.3 wt%, very close to the nominal value (ca. 5 wt%.). Cu leads to a decrease in C content in the hybrid material and to a remarkable increase in the surface pH from 3.8 up to 6.2.

Table 1. Chemical composition, surface pH and selected textural parameters the photocatalysts evaluated from N_2 adsorption isotherms at 77 K.

Sample	S_{BET} [a] ($m^2\ g^{-1}$)	V_{total} [b] ($cm^3\ g^{-1}$)	V_{micro} [c] ($cm^3\ g^{-1}$)	Ti + O [d] (wt.%)	C [d] (wt.%)	Cu [d] (wt.%)	pH [e]
TiO_2	145	0.162	0.032	79.8	20.2	---	3.8
Cu@TiO_2-C	127	0.117	0.033	78.2	17.5	4.3	6.2

[a] BET surface area; [b] Total pore volume at p/po ≅ 0.99; [c] Micropore volume, Dubinin-Raduskevich equation; [d] TiO_2, C and Cu content; [e] Surface pH.

Fig. 1A shows the N_2 adsorption-desorption isotherms at 77 K. Both materials show type IV isotherms[6], characteristic of mesoporous materials. Differences in the hysteresis loop (type H_2) indicate a wider distribution of mesopores sizes for TiO_2-C than Cu@TiO_2-C. Contribution of micropores to the total pore volume is ca. 20% and 28% for TiO_2-C and Cu@TiO_2-C, respectively. Cu promotes a slight enhancement in the micropore contribution and a decrease in C content from ca. 20.2 to 17.5 wt.%. TEM images (Fig. 1B and 1C) show the materials are constituted by micrometric spheres which distribution of sizes increased in the presence of Cu (ca. 4.0 μm and 5.2 μm) suggesting that Cu nanoparticles are intercalated within the TiO_2-C matrix. The XRD patterns (Fig. 1D) shows that anatase is the main crystalline phase of TiO_2 (ca. 90%) with a mean crystal size about 9.3 nm in both samples. In addition, XRD pattern of Cu@TiO_2-C sample presents two peaks at 43.2° and 50.6° that can be attributed to elemental Cu nanoparticles -with a mean size of ca. 9.7 nm-.

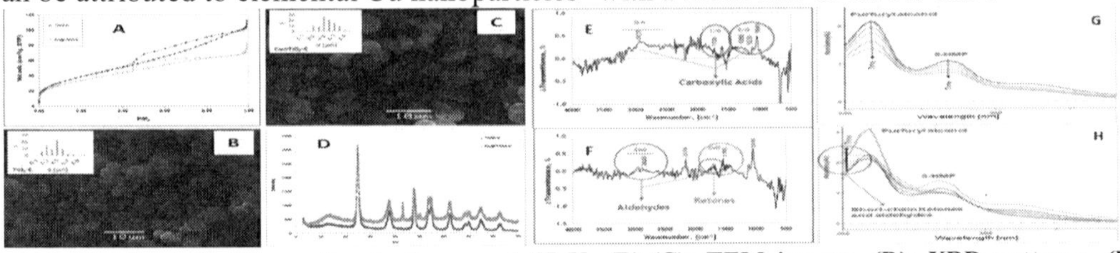

Fig 1. (A): N_2 adsorption/desorption isotherms at 77 K. (B)-(C): TEM images. (D): XRD patterns. (E)-(F): Differential transmittance from the FTIR spectra after 120 min irradiation of Gly on TiO_2-C and Cu@TiO_2-C. (G)-(H): Photocatalytic conversion of FA on TiO_2-C and Cu@TiO_2-C.

Fig. 1E and 1F show the differential transmittance data obtained from the FTIR spectra after 120 min irradiation of Gly. Carboxylic acids are formed on TiO_2-C sample while ketones and aldehydes along with carboxylic acids, are formed on Cu@TiO_2-C. These results agree with the surface pH of samples: the photooxidation of Gly to carboxylic acid occurs on TiO_2-C catalysts due to its acidic surface pH, while reduction of alcohols groups on Gly occur on the neutral Cu@TiO_2-C, where the high electron affinity of Cu is the driving force for the reduction reaction. Fig. 1G and 1H show the photoconversion of the FA and 5-HMF mixture. In absence of Cu, the absorbance peaks in the UV-vis spectrum decreased as function of time indicating that the photomineralization of the FA + 5-HMF mixture is promoted. On the contrary, the Cu-containing catalyst leads to a different reaction mechanism which is inferred by the increase in the absorbance below the main peak of FA at 217 nm and the increase for the peak below 268 nm, corresponding to 5-HMF. It may be inferred that short-chain ketones and aldehydes could be formed. A quantitative study is currently ongoing to elucidate the reaction mechanisms for the photocatalytic valorization of Gly, FA, 5-HMF.

Conclusions

Our preliminary results open new perspectives for the development of low cost and selective C-based catalysts for the photoassisted valorization of biomass′ derivatives molecules. The catalysts showed activity for the photoconversion of FA and Gly. The incorporation of Cu resulted in a catalyst with a higher selectivity for the electron-donor involved reactions.

Acknowledgment

J.Matos thank the financial support of Chilean projects: FONDECYT 1161068, Basal Program PFB-27 and Franco-Chilean network on biochars valorisation BIOCval2E (REDES-170004).

References

1. Matos J., Marino T., Molinari R., García H. (2012). *Appl. Catal. A Gen.,* 417– 418, 263.
2. Jessop P.G., Joó F., Tai C.C. (2004). Coord. Chem. Rev. 248, 2425–2442.
3. Mariscal R., Maireles-Torres P., Ojeda M., Sádaba L., López Granados M. (2016). *Energy Environ. Sci.* 9, 1144.
4. Quispe C.A.G., Coronado C.J.R., Carvalho Jr. J.A. (2013). *Renewable and Sustainable Energy Reviews* 27, 475.
5. Reddy S.N., Nanda S., Kozinski J.A. (2016). *Chemical Engineering Research and Design* 113, 17.
6. Thommes M., Kaneko K., Neimark A.V., Olivier J.P., Rodriguez-Reinoso F., Rouquerol J., Sing K.S.W. (2015). *Pure and Applied Chemistry* 87, 1051.

EMPHASIZING THE ROLE OF CARBON PORES IN ELECTROCATLYSIS

Teresa J. Bandosz*

Department of Chemistry and Biochemistry, The City College of New York, New York, NY 10031

*Presenting author's e-mail: tbandosz@ccny.cuny.edu

Introduction

Both oxygen reduction (ORR)[1] and CO_2 electroreduction ($CO_{2E}RR$)[2] are important processes addressing current energy needs and environmental protection concerns. To further address the latter, recently there has been reported an extensive search for non- noble metal and metal -free efficient catalyst for this processes. Among those ,the most promising candidates are carbon-based materials. So far, mainly nonporous nanoforms of carbons such as CNT or graphene has been explored, either doped with heteroatoms (nitrogen, sulfur, boron) or modified by an introduction of defects. Surface dopants and defects in the carbon matrix were identified as catalytic centers of these processes.
Recently, based on the performance results of modified graphene and carbon nanotubes we have explored the application of nanoporous carbons as ORR[3,4] and CO_2ERR catalysts[5-7] . An enhancement in the performance owing to the porosity involvement in the overall reduction processes was our goal. Therefore, the objective of this minireview is too indicate the role of adsorption forces, and especially those in small pores of strong adsorption potential in ORR and CO_2ERR processes. The conclusions presented here are based on the study of the catalytic performance of various porous carbon materials, either heteroatom-free or doped with N and S and on the extensive analyses of the surface features of the initial and spent catalysts.

Materials and Methods

Materials tested were either carbon foams, home-made polymer based carbons, or commercial carbons modified with either GO or with S, N or N and S doping. The initial and spent catalyst were extensively characterized using adsorption of nitrogen (porosity), XPS, potentiometric titration, TA/MS, FTIR, HRTEM, HRSEM, EDX, ICP, and conductivity measurements.
The ORR performance was tested using RRDE in 0.1 M KOH. From the experiments the onset potentials, numbers of electron transfer, %of H_2O_2, and kinetic current density were calculated.
CO_2ERR experiments were run in a home-designed three electrode cell in a 0.5 M $KHCO_3$ electrolyte saturated with CO_2. The products of the reactions were identified using GC and from the results the Faradaic efficiencies (FE) for the products formation, kinetic currents and onset potentials were determined. In both approaches, the stability of the catalysts was tested.

Results and Discussion

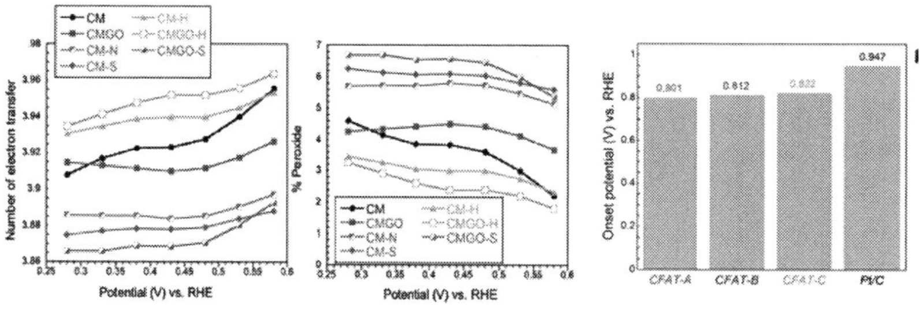

Figure 1. Number of electron transfer (left), percent of H_2O_2 formed on heteroatom free and S and N modified carbons, and the onset potentials on heteroatom free-carbons. Adapted with permission from Ref.[3, 4]. Copyright 2016 and 2017, American Chemical Society

The examples of the results obtained in the case of ORR are presented in **Figure 1**. On the materials tested the number of electron transfer was very close to 4, and the onset potential, although smaller than that for Pt/C depended on the size of ultramicropores, for carbon free of heteroatoms. For the carbons,which were modified with S or N heteroatoms, and had smaller sizes and volume of ultramicropores, the performance was worse than those of carbon rich in ultramicropores

The enhancement in the performance,due to ultramicropores, was linked to the withdrawal of oxygen from water by adsorption forces in small pores and by their hydrophobicity, simultaneous acceptance of electrons by split oxygen atoms and their protonation. Affinity of OH- to water present at the pore entrance freed the oxygen adsorption centers for further reduction processes (**Figure 2**).

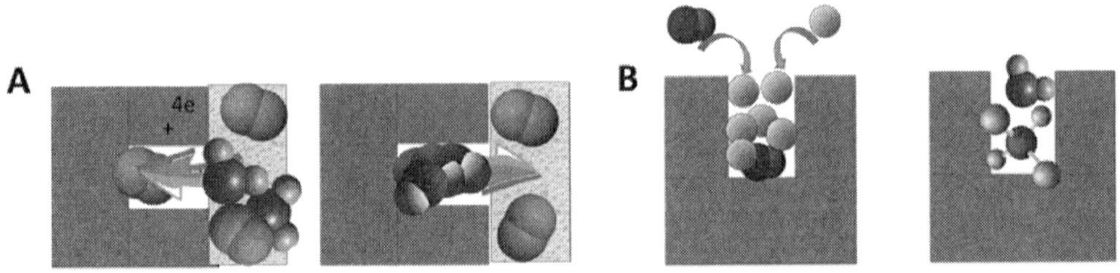

Figure 2. Role of pores in ORR (A) and on CO_2ERR (B).

In the case CO_2 ERR, the FE for CO formation on porous carbon modified with melamine reached 40 % with 40 % FE for H_2. The strong adsorption of CO_2 on positively charge surface centers (pyridic nitrogen) was indicated as promoting the reduction process at relatively low overpotential. Interestingly, CH_4 and other hydrocarbons (C_3) were also detected with small FEs. Formation of methane is very unique and was detected only on porous carbon. Its apperance was linked the possibility of pseudo Fischer -Tropsh reaction in ultramicropores/nanoreactor where, owing to strong adsorption of CO, high pressure and presence of hydrogen from water reduction, CH_4 could be formed.

Conclusions

The results collected suggest that the performance of carbon catalyst in ORR and CO_2ERR can be improved by the involvement of porosity. The latter, combined with heteroatom-based catalytic centers can lead to the performance comparable to that Pt/C in ORR or to syngas and CH_4 formation in CO_2ERR.

Acknowledgment

The contributions of Drs. W. Li, A. Clezard, V. Fierro, E. Rodriguez-Castellon, M. Seredych, H. He, N. Fechler, A. Szczurek, N. Travlou, K.Savaram, M. Szczurek, and J. Encalada, Q. Li, C. Delgado-Sanchez to this research are acknowledged.

References

1. Shao, M., Chang, Q., Dodelet, J,.P., Chenitz, R. (2016) Recent Advances in Electrocatalysts for Oxygen Reduction Reaction. Chem. Rev. 116, 3594–3657
2. Zhang, J.; Xia, Z., Dai, L. (2015) Carbon-based electrocatalysts for advanced energy conversion and storage. Sci. Adv.1:e1500564
3. Seredych, M., Szczurek, A., Fierro, V., Celzard, A, Bandosz, T.J. (2016) Electrochemical Reduction of Oxygen on Hydrophobic Ultramicroporous PolyHIPE Carbon. *ACS Catal. 6*, 5618–5628.
4. Encalada, J., Savaram, K., Travlou, N.A., Li, W.; Li, Q.,Delgado-Sánchez,C., Fierro, V., Celzard, A., He, H., Bandosz T.J. (2017) Combined Effect of Porosity and Surface Chemistry on the Electrochemical Reduction of Oxygen on Cellular Vitreous Carbon Foam Catalysts,*ACS Catal.* 7 , 7466–7478.
5. Li, W., Seredych, M., Rodriguez-Castellon, E.,Bandosz, T.J.. (2016) Metal Free Nanoporous Carbon as a Catalysts for Electrochemical Reduction of CO_2 to CO and CH_4. *ChemSusChem* 9, 606-616.
6. Lu, W., Herkt, B., Serdych, M., Bandosz, T.J. (2017) Pyridinic-N groups and ultramicropore nanoreactors enhance CO_2 electrochemical reduction on porous carbon catalysts. *Appl, Catal.* B 207, 195-206.
7. Li, *W., Fechler,N.;* Bandosz, T.J.(2018) Chemically heterogeneous nitrogen sites of various reactivity in porous carbons provide high stability of CO_2 electroreduction catalysts. *Appl. Catal. B* 234, 1-9

Carbon2018

The study toward to the realization of sustainable CFRTP having both high mechanical properties and recyclability

Toshihira Irisawa[1*], Ryo Hashimoto[1], Naoaki Nagao[1], Masahiro Arai, Yasuhiro Tanabe[1]

[1]*Graduate School of Enginieering, Nagoya University, Nagoya, Japan*

*Presenting author's e-mail: irisawa.toshihira@material.nagoya-u.ac.jp

Introduction

Carbon fiber reinforced thermoplastics (CFRTPs) have been focused for their high production rate compared with traditional CFRPs. Some commodity thermoplastics have been chosen as the matrix in terms of cost. If the CFRTP can be obtained with good properties, CFRTP will be used for automotive application and CFRTPs contribute to the realization of sustainable society in terms of energy saving and decreasing of CO_2. However, it is also concerned that these CFRTPs are insufficient in terms of thermal stability even for automotive application.

Some aromatic thermoplastics are known as polymer having high thermal stability. Therefore, this study deals with CFRTPs made with polyethersulfone (PES) which is an aromatic thermoplastic as the next generation CFRTPs for realization of sustainable society. The research about these CFRTPs are few, hence the possibility of these CFRTPs was investigated in terms of the mechanical properties and thermal stability. Accordingly, it is better that CFRTPs themselves become the sustainable material, if the realization of sustainable society is hoped. Therefore, this study also focuses on the recyclability of these CFRTPs. As recycle methods, both pyrolytic and chemical processes which are methods to separate and recover CFs from CFRPs has been compared in terms of the tensile properties of CFs.

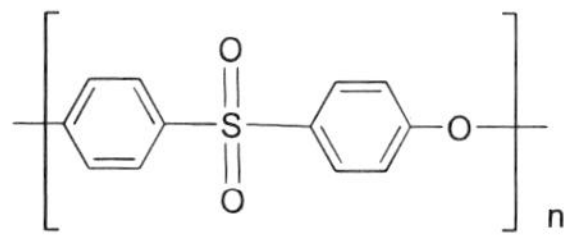

Figure 1. The molecular structure of PES

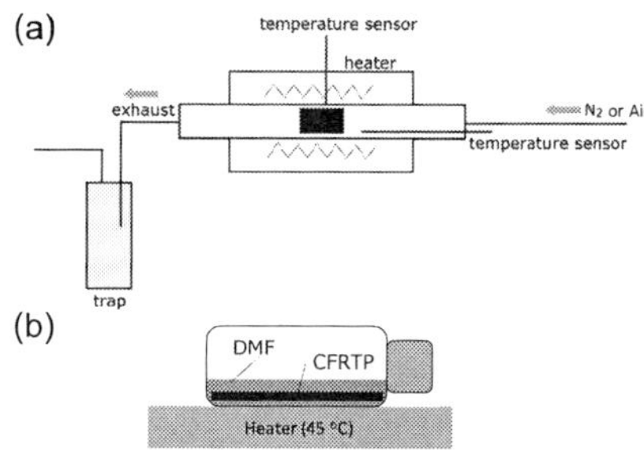

Figure 2. Schematic illustration of (a) pyrolytic method and (b) depolymerisation under ordinary pressure method to recover the recycled CFs.

Materials and Methods

PAN-based CF (HTS-40-3K, Toho Tenax Co., Ltd.) plain fabrics were used as the reinforcement fibers and PES (**Figure 1.**) with T_g of about 220 °C were used as the matrix polymer, and Prepregs impregnated with PES (resin Contents of 37 wt%) were supplied by Toho Tenax Co., Ltd. CFRTPs were molded under the best condition for us at the present stage (at 320 °C) by hot-press [1]. Mechanical properties of CFRTPs were evaluated by a 3-point bending test at -40 to 200 °C.

For the investigation of recyclability of these CFRTPs, Re-CFs (Re-pyro-CF and Re-chem-CFs) were recovered by using the pyrolytic process and chemical process (**Figure 2.**) under the best condition in each process for us at the present stage. The CFRTP was pyrolyzed at 550 °C for 30min in the air for the pyrolysis process, and it was dissolved in N,N-dimethylformamide as a solvent at 45 °C for 4 h for the chemical process. The mechanical properties of fresh CFs (Fre-CFs) and each Re-CFs were measured by the single fiber tensile tests (ISO-11566). For each type of CFs, 50 single fibers were tested. The density of each type of CFs was measured with the sink-float method and the WAXD analysis was carried out for evaluation of crystallinity as interlayer spacing (d_{002}) and the carbon layer stack height (L_c).

Results and Discussion

The results of bending tests of CFRTP at a target temperature from -40 to 200 °C shows in Figure 3. The average modulus and strength at 25 °C achieved over 85% and 80% of theoretical values calculated from role of mixture, respectively. It can be said that these value means a top quality as CFRTPs.

The modulus showed almost constant below 150 °C. In addition, the strength still is over 450 MPa at 150 °C, and it is thought that these values are sufficiently high as structural materials. Therefore, these results revealed that these CFRTPs are promising materials in the use until 150 °C. On the other hand, the modulus immediately decreased at over 150 °C,and the strength at 100 °C also became 70 % of it at R. T. The thermal stability of these CFRTPs might require further improvement to expand the application areas. The tensile properties and the structural parameters of each CF are shown in **Table 1**, respectively. Tensile modulus of each Re-CFs was almost constant compared with Fre-CFs. It is known that the tensile modulus of CFs correlate with the structural parameters of them [2]. In this case, pyrolytic and chemical processes did not significantly affect the density and structural parameters of each Re-CFs. Therefore, the tensile modulus of each Re-CF was same or similar to that of Fre-CFs.

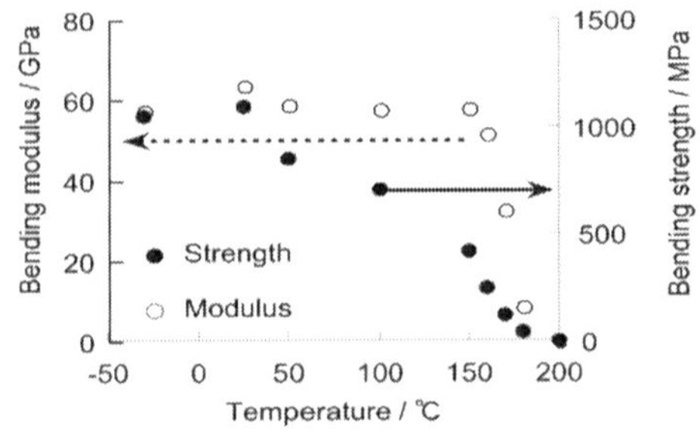

Figure 3. The bending strength and modulus of CFRTP at the target temperature.

Table 1. The tensile properties, structural parameters of crystallites and density of Fre-CF and each Re-CFs.

	Tensile test		WAXD		Sink-float
Type of CFs	Modulus (GPa)	Strength (MPa)	d_{002} (nm)	L_c (nm)	Density (g/cm^3)
Fre-CF	213± 14	3.99±0.72	0.354	1.43	1.79
Re-pyro-CF	208± 12	2.31±0.53	0.355	1.38	1.78
Re-chem-CF	211± 20	4.01±0.75	0.352	1.47	1.79

On the other hand, the tensile strength of Re-pyro-CFs decreases more than 40% compared to Fre-CFs, while that of Re-chem-CFs did not decrease. In pyrolytic process, to remove matrix resin from CFRTPs as much as possible, CFRTPs needed to be exposed under a severe condition at higher temperature or for longer time in the Air. Although CFs certainly have high thermal stability and the pyrolytic treatment is only for removing matrix resin, the pyrolytic process cause the damages according to oxidation decomposition of CF's surface. As a result, the strength of Re-pyro-CFs decreased. In chemical process, Re-chem-CFs could be recovered without a severe condition like a pyrolytic process. Therefore, the tensile strength was about same as that of Fre-CFs.

Conclusions

In this study, the possibility of CFRTPs made with PES as the sustainable material was investigated in terms of the mechanical properties, thermal stability and recyclability. This study reveals that the mechanical properties of these CFRTP are sufficiently high as structural materials and are promising materials in the use up to 150 °C. Furthermore, it is revealed that these CFRTPs also have high recyclability. Although the tensile strength of Re-pyro-CFs were decreased, Re-chem-CFs of these CFRTPs could be recovered without the decreasing of tensile strength.

Ttherefore, it is thought that these CFRTPs have the possibility of the extensive application. I think that this fact will contribute to the realization of sustainable society.

Acknowledgment

This work was supported by CSTI, SIP Program, "Development of High-Rate Production Innovative Aircraft PMC Products and Quality Assurance Technology" (Funding: JST). The materials are supplied by Toho Tenax co., Ltd.

References

1. T. Irisawa, *Carbon 2018*, poster No. 448
2. DD. Edie, Carbon, **36(4)**, 345 (1998).

CARBON MATERIALS AS CATALYST SUPPORTS AND CO-CATALYSTS FOR BIOMASS UPGRADING PROCESSES

Salvador Ordóñez*, Eva Díaz, Laura Faba, Diego Garcés
Department of Chemical & Environmental Engineering, Faculty of Chemistry, University of Oviedo, Spain
*Presenting author's e-mail: sordonez@uniovi.es

Introduction

Most of the strategies for the upgrading of biomass-derived platform molecules involve multi-functional catalysts. Furthermore, many of these reactions take place in biphasic systems, with a marked influence of the mass transference between phases. The use of carbon materials can be very useful for tuning catalyst properties or to promote interphase mass-transfer effects. We consider the upgrading route cellulose to biogenic aldehydes (furfural and 5-HMF) with further transformation of these aldehydes into long chain hydrocarbons and chemicals via aldol condensation and hydrogenation. The first step is performed at mild conditions using homogeneous acid as catalysts, and the process is limited by the low water solubility of the aldehydes. We demonstrate that the use of carbonaceous materials as co-catalysts enhances the mass transfer of the aldehydes from aqueous to organic phase, improving reaction rates and limiting further non-desired decompositions. Concerning to the aldol condensation, we demonstrate that supporting the Mg-Zr mixed oxide (typical catalyst for this reaction) on non-microporous carbonaceous materials, such as carbon nanofibers (CNFs) or high surface area graphites (HSAGs) leads to a better performance, justified by a more appropriate basic sites distribution and the corresponding interactions between reactants and the support. This interaction also improves the behaviour of the catalysts used for the hydrogenation (HDO) of these adducts. The use of carbon materials, especially HSAGs, has important advantages in comparison to common inorganic supports (hindering the formation of carbonaceous deposits). By contrast, active carbons (AC) lead to poorer results, mainly caused by mass transfer limitations.

Materials and Methods

All the reactions were carried out in a batch Autoclave reactor, pressurized up to 10 bar and under stirring. For the aldehydes production, a biphasic system 1:1 water to MIBK was used (175 mL each) feeding 5.83 g of cellulose. A co-catalytic system was considered, combining 200 ppm of HCl (37 %) and different amounts of activated carbon or carbon nanotubes (T = 140 °C, t = 24 h). The aldol condensation was studied in aqueous phase (250 mL), at 50 °C, using a 5 % of organic compounds (furfural/acetone molar ratio 1:1) and supporting the active phase (Mg-Zr) on different carbonaceous supports (HSAG100, HSAG300, HSAG500 and CNF) combining incipient wetness impregnation and co-precipitation methods. Finally, HDO tests studied supporting 1 % Pt on carbon materials (HSAG300, CNF), using hexane or acetone as solvents (0.25 L) with 0.42 g of reactants (C8 or C13 condensation adducts obtained in the previous step), with 25 bar of H_2 and 220 °C.

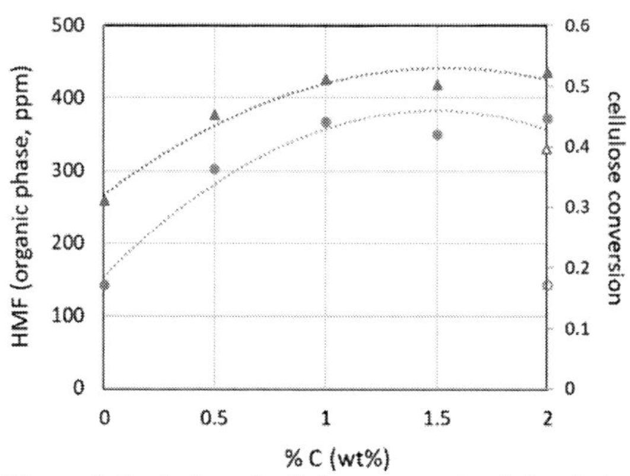

Figure 1: Evolution of cellulose conversion (triangles), and HMF concentration in the organic phase (circles) in the bifasic cellulose hydrolysis promoted with carbon materials (closed symbols: CNTs; open symbols: AC).

Results and Discussion

As observed in the Fig. 1, the presence of CNTs in the reaction media, largely improves both the cellulose conversion and the selectivity to, HMF which is recovered in the organic phase. This effect is not observed when using similar amounts of activated carbon. These results suggest that CNT act as a phase transfer catalysis, adsorbing HMF from the aqueous phase and releasing it into a organic phase. As several reaction steps are reversible, this transport enhances the overall cellulose conversion. Concerning the aldol condensation, preliminary results determined that the highest activity was obtained with Mg-Zr mixed oxide, with 81 and 54 % of furfural and acetone conversion, respectively, after 24 h, with a final selectivity of 15 and 62 % for C8 and C13 adducts. In these experiments, an increase of catalytic loading was not directly related to an increase in the activity, suggesting a poor exposition of active sites. In order to corroborate this hypothesis, Mg-Zr was supported on different mesoporous carbon materials (HSAG, CNF) and reactions were repeated with the same amount of catalyst (a high decrease in the total concentration of active acid/basic sites, but a high dispersion, because of the support). Results with CNF discarded this material because the lack on anchoring sites on CNF hinders the deposition of the active phase, obtaining poor results. On the contrary, there was a very good interaction between the mixed oxide and the HSAG, mainly when these catalysts were prepared by co-precipitation, obtaining a regular and well dispersed deposition. As consequence, the furfural conversion increases up to 96.5 %, with a global selectivity of 80-90 % for both products after only 4 h, obtaining kinetic constants more than 3 times higher than those previously estimated.[1] If these results are normalized by basic site concentration, this ratio increases to 80, as it is observed in Figure 2a. In addition, a high increase in the stability was observed, in contrast to the fast deactivation observed with the unsupported Mg-Al. The influence of carbon materials was also tested in the last step of the proposing process: the hydrodeoxygenation of condensation adducts to obtain the alkanes. For this reaction, a good equilibrium between metal and acid sites is required, leading to a key role of the support properties and dispersion capacity. Acid supports such as alumina, zeolites and AC presented a high activity, with some problems related to side reactions and irreversible adsorption, mainly inside the micropores. The highest n-octane productivity (close to 30 %) was obtained with Pt/HSAG, offsetting its low acidity by a high metal dispersion (Fig. 2b). In addition, side reactions were prevented and the adsorption was limited. Considering these results, final experiments were carried out analysing a possible one-pot configuration. More than 12 % of alkanes were obtained from furfural and acetone, with a mechanical mixture of Mg-Zr/HSAG + Pt/Al_2O_3, with a high selectivity of n-tridecane.[2]

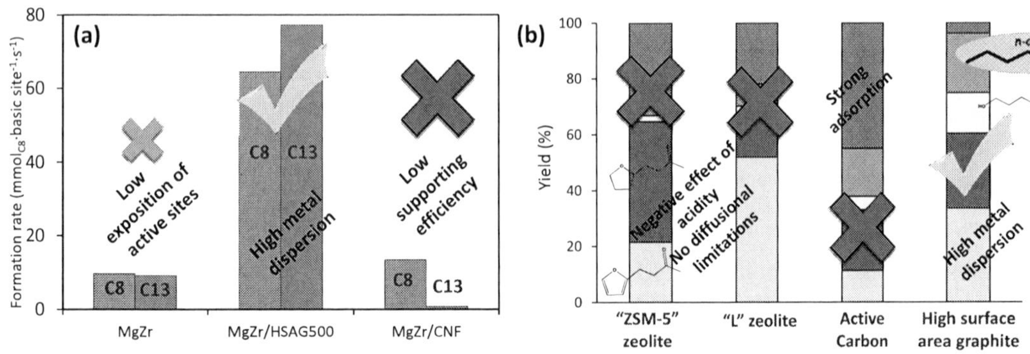

Figure 2: (a) Comparison of formation rates obtained during the aldol condensation. (b) Comparison of main activity results during the C8 hydrodeoxygenation obtained by supporting Pt nanoparticles on different supports

References

1. Faba, L, Díaz E., Ordóñez, S. (2013). Improvement on the catalytic performance of Mg-Zr mixed oxides for furfural-acetone aldol condensation by supporting on mesoporous carbons. *ChemSusChem* 6. 463-473.
2. Faba, L, Díaz E., Ordóñez, S. (2014). One-pot aldol condensation and hydrodeoxygenation of biomass-derived carbonyl compounds for biodiesel synthesis. *ChemSusChem* 7, 2816-2820.

HIGH-VOLTAGE DURABILITY AND DEGRADATION MECHANISM OF SEAMLESS ACTIVATED-CARBON ELECTRODE FOR ELECTROCHEMICAL CAPACITORS

Soshi Shiraishi[1*], Tomomi Tagaya[1], Ryohei Oyama[1], Yoshikiyo Hatakeyama[1], Hidehiko Tsukada[2]

[1]*Graduate School of Science and Technology. Gunma Univ., Kiryu, Japan*
[2]*AION Co. Ltd., Koga, Japan*
*soshishiraishi3@gunma-u.ac.jp

Introduction

The electric double layer capacitor (EDLC) is an electrochemical energy-storage device with high power density. In the EDLC, the higher voltage operation than 3 V is required for improving the energy density and achieving more reliability, but electrochemical decomposition during high voltage charge causes the decrease in the capacitance and the increase in the internal resistance through i) the micropore blocking by the decomposition products, ii) the destroy of capacitor cell by the gas evolution, or iii) the breakdown of electric network in the electrode. The authors are focusing on the suppression of the breakdown of electric network.

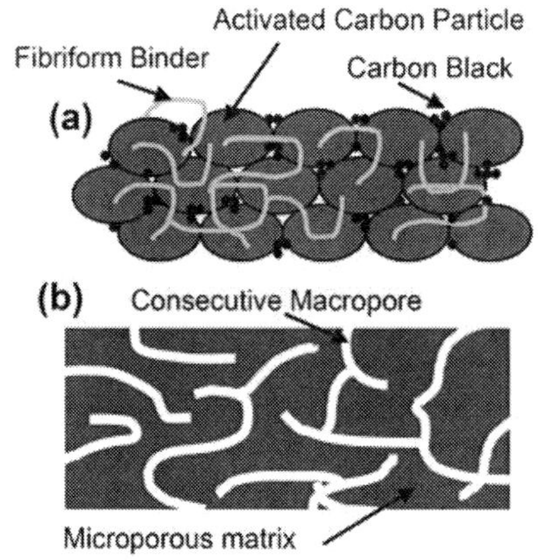

Fig. 1 Schematic model of (a) conventional and (b) seamless activated carbon electrode for EDLC.

The conventional EDLC electrode, which is composed of the activated carbon powder, carbon black, and binder (**Fig.1**(a)), is easily deteriorated by the breakdown of electric network at high voltage charge. Thus, the author's newly proposed the seamless structure of the activated carbon electrode containing no particle-boundaries (**Fig.1**(b)) to realize high durability against higher voltage charge for the EDLC [1]. **Fig.2** shows the Ragone plot of the EDLC using seamless activated carbon electrode before and after the floating durability test with high voltage charge (3.5 V, 100 h, 70 C). The excellent durability, which is demonstrated using seamless activated carbon electrode, is derived from the electrochemical stability against the breakdown of electric network in the electrode.

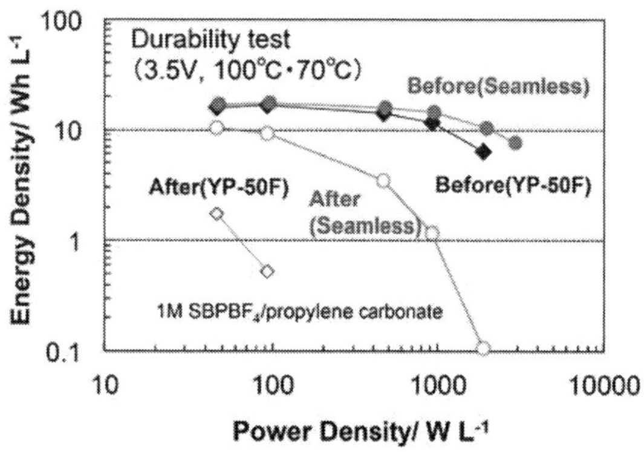

Fig. 2 Ragone plot of EDLC using densified seamless activated carbon electrode or conventional composite electrode (YP50F) before and after durability test (3.5 V, 100 h, 70℃).

The seamless activated carbon electrode has other advantages to conventional activated carbon electrode in addition to the high durability. One of them is related to the facile characterization of the electrode. For example, the analytical results for the tested electrode can be discussed without any consideration of the conductive additives and binder influence. Therefore, it is useful to understand the degradation mechanism using the seamless activated carbon electrode. In this paper, the authors discuss the degradation mechanism of the EDLC at high-voltage charge with the analysis results using seamless activated carbon electrode.

Materials and Methods

The seamless activated carbon electrode (MLAC) was prepared by carbonizing, and CO_2-activating (900 °C) the consecutively macroporous phenolic resin matrix (MICROLITE, AION). **Fig.3** shows SEM image of the MLAC. The two-electrode cell was assembled by the MLAC electrodes and organic electrolyte (1M $(C_2H_5)_3CH_3NBF_4$ / propylene carbonate). The capacitance was measured by the galvanostatic method (80 mA g^{-1}, 0~2.5 V) at 40 °C. The durability test was conducted by the floating condition (3.5 V, 100 h, 70 °C).

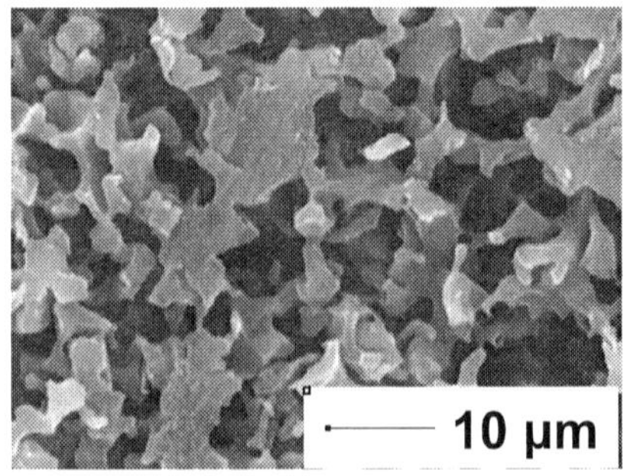

Fig. 3 Scanning electron microscopic image (SEM) of MLAC.

Results and Discussion

The results of the durability test were summarized in **Table 1**. The capacitance retention of the MLAC2000 (80 %) is much higher than the MLAC1300 (22 %), while there are almost no differences in the integrated leakage current (the extent of electrochemical decomposition) between them. It indicates that ill-developed micropore is easier blocked than well-developed one under the same amount of decomposition products.

Table 1 Pore structure and high-voltage charge durability of MLACs

Electrode	S_{BET} [$m^2 g^{-1}$]	w_{micro} [nm]	Initical Capacitance [F g^{-1}]	Capacitance Retention [%]	Integrated Leacage Current [mA h]
MLAC1300	1320	0.75	28	22	30
MLAC2000	1950	0.97	33	80	32

S_{BET}: BET specific surface area, w_{micro}: micropore width (DR method)

In the both cases of the negative and positive electrode, the micropore volume is decreased by the durability test, as shown in the pore size distribution curves of **Fig.4**(a) and (b). The decrease in the micropore volume is due to the decomposition product depositing. However, the micropore-volume ratio after the durability test for the MLAC1300 is not very different from that for the MLAC2000, which is inconsistent with the capacitance retention. Moreover, the center of the pore width distribution peak is not much shifted by the durability test. This suggests that the decomposition products deposit ununiformly on the electrode (ex. on the entrance or in the path of the micropore).

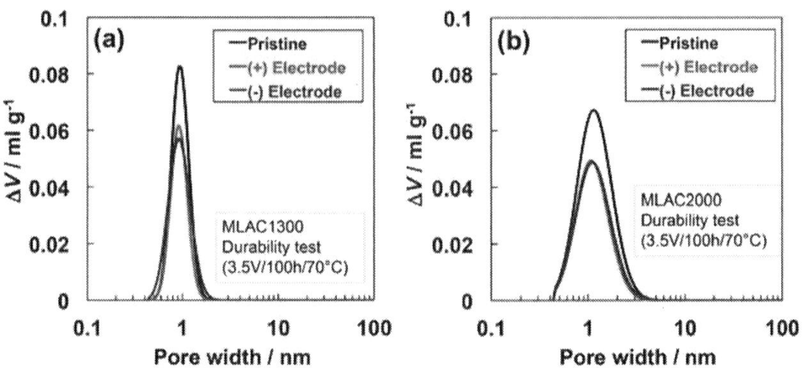

Fig. 4 Pore size distribution curves (by NLDFT) of the N_2 adsorption isotherms of the (a) MLAC1300 or (b) MLAC2000.

Conclusions

The seamless activated carbon electrode is very useful to discuss the degradation mechanism of the EDLC. In this paper, it can be concluded that the capacitance decline by the high-voltage charge is due to the local deposition of electrochemical-decomposition in the electrode.

Acknowledgment

A part of this research is supported by JSPS KAKENHI Grant No. 17H03123.

References

1. S. Shiraishi. (2013). Highly-Durable Carbon Electrode for Electrochemical Capacitors. *Boletín del Grupo Español del Carbón* 28, 18-24.

Nanographenes by Solvent-free Syntheses - The Mechanochemical Scholl Reaction

Sven Grätz[1*], Doreen Beyer[2], Valeriya Tkachova[2], Sarah Hellmann[1], Reinhard Berger[2], Xinliang Feng[2] and Lars Borchardt[1]

[1] *TU Dresden: Chair of Inorganic Chemistry, Dresden, Germany*
[2] *TU Dresden: Chair for Molecular Functional Materials, Dresden, Germany*

*Presenting author's e-mail: sven.graetz@tu-dresden.de

Introduction

Nanographenes and graphene nanoribbons are promising materials with potential applications in organic and nano-electronics.[1] During the bottom-up synthesis of these nanographenes the Scholl reaction[2,3], a Lewis-acid catalysed oxidative cyclodehydrogenation is utilized to planarize precursor systems by forming aryl-aryl bonds and creating a fully pi-conjugated material. Especially for the larger precursors solubility is a main concern but even the smallest planarized hexabenzocoronene (HBC) is basically insoluble in common organic solvents. The common workaround is the introduction of solubilizing groups, resulting in additional synthesis steps. Thereby, the atom economy of the process suffers, hindering the scale-up needed for industrial applications. In the meantime, the search for the ideal, environmental friendly and preferably cheap solvents has been a persistent topic in general chemistry. But where solvent based chemistry fails other methods have to step in. In recent years mechanochemical and other solid-state methodologies have been explored as a powerful tool, offering at least as much flexibility as solvent-based processes.[4,5]

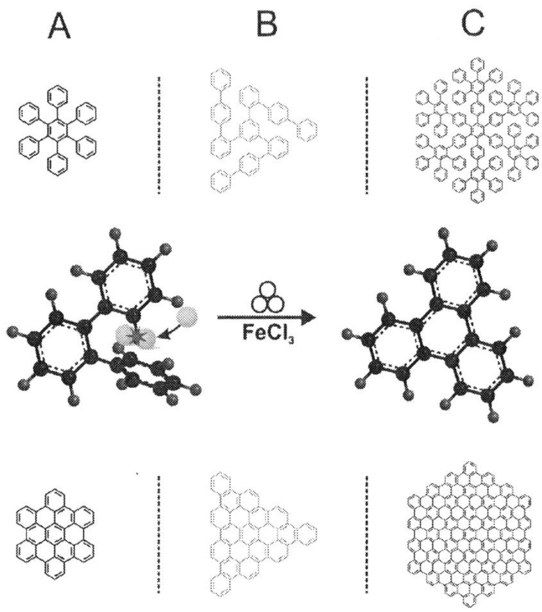

Figure 1. Scholl reaction of different precursors (C42H26 (A); C60H42 (B); C222H150 (C)) with iron(III) chloride in a planetary ball mill to produce the benchmark PAHs HBC, triangular-shaped C60 and C222.

Materials and Methods

Starting from hexaphenylbenzene (HPB), a commercially available precursor, we investigated the feasibility of the Scholl reaction under mechanochemical conditions in a planetary ball mill (PBM). After initial experiments turned out to be very promising, we could quickly raise the yield to 95% in a 45 ml zirconium dioxide milling vessel with 10mm balls at 800 rpm adapting the literature known conditions for solution synthesis, of 12 eq. FeCl3 per H atom involved in aryl-aryl bond formation (**Table 1**). The experiments were conducted with 0.1 g of the HPB and NaCl as bulking material. The quality of produced nanographenes has been confirmed by MALDI-TOF mass spectrometry and UV-Vis absorption spectroscopy.[6]

Results and Discussion

In this study, we focused on developing a protocol for the solvent-free Scholl reaction under mechanochemical conditions as a versatile tool for the cyclodehydrogenation, "graphitization" of oligophenylene precursors into benchmark nanographenes such as HBC, triangular shaped C60H42 and C222 (**Fig. 1**). MALDI-TOF mass spectrometry of HBC-1 showed the most intense signal at m/z = 522 g mol^{-1}, which represents exactly the target molecule (**Fig. 2A**). The purity of the sample was further confirmed using UV/Vis spectroscopy. After this preliminary result, we employed a design of experiments approach to handle the vast amount of parameters present in a mechanochemical reaction. This investigation confirmed that indeed all of these parameters influence the reaction while in general a higher energy input favours a high yield of the reaction. In an additional trial we demonstrated, that an increase of milling time consequently leads to a higher percentage of chlorinated side products. It is therefore of key interest to track the reaction and determine its end in order to obtain the cleanest product possible. Therefore, we reverted to the in-situ pressure measurement of the milling vessel, to elucidate the reaction further. Applied to our system, we can observe a step rise of pressure in the first minutes of the reaction (**Fig. 2B**). Caused by the release of HCl during the reaction, this rise occurs in different slopes depending on the milling speed.[6]

Table 1. Reaction conditions and yields

Sample	Milling material [-]	Milling time [min]	Yield [%]
HBC-1	ZrO_2	60 min	95 %
HBC-5	ZrO_2	30 min	92 %
C60	ZrO_2	60 min	81 %
C222	ZrO_2	60 min	89 %

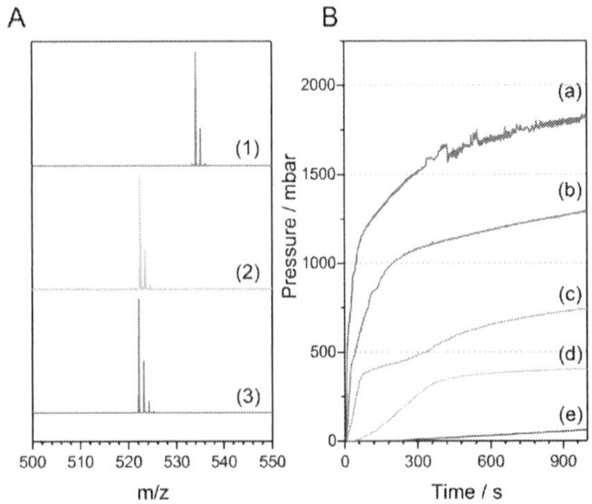

Figure 2. A: Scholl reaction of C60H42 to C60H24, milled with zirconium dioxide recorded MALDI-TOF spectra of HPB (1), HBC (2), and calculated MALDI-TOF spectra of HBC (3) B: Development of the vessel pressure during the milling process of HPB for different milling speeds. (a) 800 rpm, (b) 600 rpm, (c) 400 rpm and (d) 200 rpm (e) 100 rpm.

Conclusions

In summary, we described a novel and innovative mechanochemical process for the Scholl reaction of benchmark nanographenes. Using $FeCl_3$ in a solvent-free protocol in a planetary ball mill, we could demonstrate that these reactions proceed in as little as 30 minutes and are not limited by the solubility of the PAH. In addition, an in-situ study of pressure and temperature during the milling process was conducted. This new reaction route paves the way towards larger extended nanographenes and graphene nanoribbons and renders the introduction of solubilizing groups obsolete.

Acknowledgment

We gratefully acknowledge the Federal Ministry of Education and Research (Bundesministerium für Bildung und Forschung, BMBF) for support of the Mechanocarb project (award number 03SF0498). This research was supported financially by ERC grants on 2DMATER, the EC under Graphene Flagship (No. CNECT-ICT604391), the Center for Advancing Electronics Dresden (cfaed), the European Social Fund, and the Federal State of Saxony (ESF-Project "GRAPHD", TU Dresden). The authors acknowledge Sebastian Ehrling for the SEM/EDX measurements and Dr. Ajayakumar Murugan Rathamony for synthetic support.

References

1. A. Narita, X.-Y. Wang, X. Feng and K. Müllen, Chem. Soc. Rev., 44 (2015), 6616–6643.
2. R. Scholl, C. Seer and R. Weitzenböck, Ber. Dtsch. Chem. Ges, 43 (1910), 2202–2209.
3. M. Grzybowski, K. Skonieczny, H. Butenschön and D. T. Gryko, Angew. Chemie Int. Ed., 52 (2013), 9900–9930
4. S. L. James, C. J. Adams, C. Bolm, D. Braga, et al., Chem. Soc. Rev., 41 (2012), 413–447.
5. J. G. Hernández and C. Bolm, J. Org. Chem., 82 (2017), 4007–4019.
6. S. Grätz, D. Beyer, V. Tkachova, R. F. J. Berger, S. Hellmann, X. Feng and L. Borchardt, Chem. Commun., DOI:10.1039/C8CC01993B.

CO_2 CAPTURE BY CHEMICAL LOOPING PROCESSES USING C-FUELS

F. García-Labiano*, M.T. Izquierdo, L.F. de Diego, P. Gayán, A. Abad, T. Mendiara, J. Adánez

Instituto de Carboquímica, ICB-CSIC. c/Miguel Luesma Castán, 4. 50018 Zaragoza. Spain

*Presenting author's e-mail: glabiano@icb.csic.es

According to the IPCC[1], warming of the climate system is unequivocal and this is mainly consequence of the increase in the CO_2 concentration in the atmosphere. In addition, "Human influence on the climate system is clear, and recent anthropogenic emissions of greenhouse gases are the highest in history". These emissions are mainly consequence of the burning of C-fuels, mostly fossil fuels, which result in a net increase in carbon concentration in the atmosphere, soil and oceans. In fact, the natural carbon cycle in the earth is being affected. Limiting climate change would require substantial and sustained reductions in greenhouse gas emissions. The measures considered for the reduction of CO_2 emissions to the atmosphere include the improvement in the use and generation of energy, the use of nuclear energy, the boost of energy generation from renewable sources and the CO_2 capture and storage (CCS).[2]

Recently, the Paris Agreement approved a more ambitious target about CO_2 emissions urging to hold the global average temperature increase to well below 2 ºC by 2100, and to pursue efforts to limit the temperature increase to 1.5 ºC above preindustrial levels. According to the IPCC[3] predictions, CO_2 emissions should peak by 2020 and then decrease to negative values by the end of the century. Among the different options, most low-carbon scenarios rely on the integration of CCS technologies with the use of renewable C-fuels, which it is known as Bio-Energy with CO_2 Capture and Storage (BECCS). In fact, BECCS arises as the unique solution able to reach negative emissions and at the same time to obtain energy. Figure 1 shows the carbon emissions balance for the different concepts in energy generation from different C-fuels. Obviously, the implementation at high scale of CCS technology is necessary and within it the development of advanced CO_2 capture processes with reduced costs is highly encouraged. In this sense, Chemical Looping (CL) is a second generation CO_2 capture technology able to obtain high purity CO_2 streams at low cost and energy penalties for different applications including combustion for heat&electricity generation or syngas/H_2 production.[4,5]

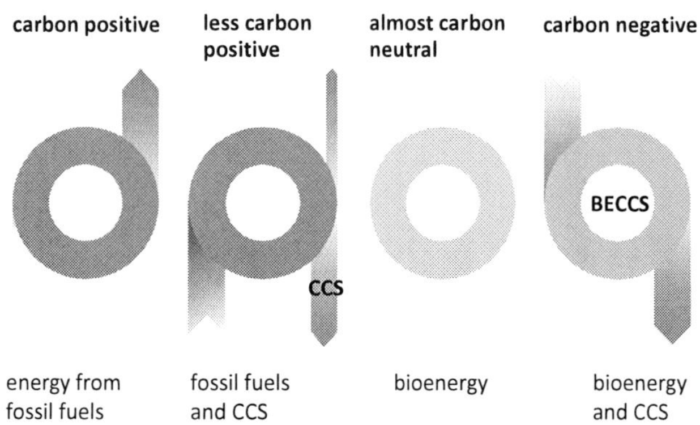

Figure 1. Carbon emission balance for different C-fuels and technologies. Adapted from IPCC.

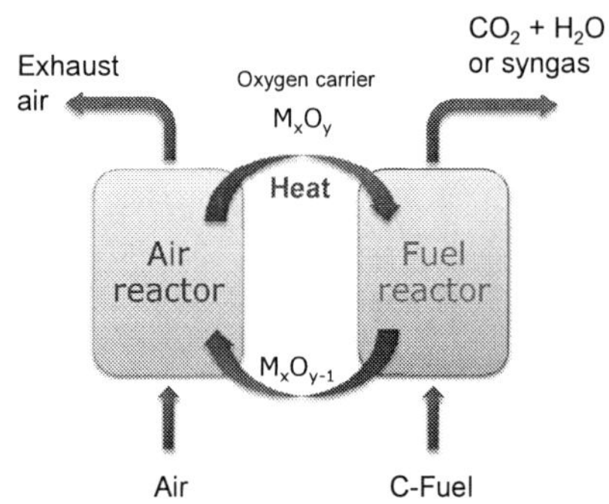

Figure 2. Chemical Looping concept.

Chemical Looping technologies include those cycling processes that use a solid material as oxygen carrier containing the oxygen required for the conversion of the fuel (fuel reactor), see Figure 2. The depleted oxygen carrier material is reoxidized before to start a new cycle in another reactor (air reactor). The technology is highly versatile since can be used with C-fuels of different state (gaseous, liquids and solid), origin (fossil or renewable), and final purposes including the combustion for heat&electricity generation (Chemical Looping Combustion -CLC-; in situ gasification CLC, iG-CLC; Chemical looping with oxygen uncoupling, -CLOU-), or the syngas/hydrogen production (Chemical Looping Reforming, -CLR-; Chemical Looping Gasification, -CLG-; Chemical Looping coupled with Water Splitting, -CLWS-).

This work summarizes the state of the art of the Chemical Looping technology, including the development of oxygen carriers, operation in existing facilities, and future research and prospects for the different processes. It also includes a techno-economic evaluation of the CL process in comparison with other competing CO_2 capture processes.

The CL technology has gained a lot of experience during last 20 years, especially for the use of fossil fuels. Currently there are about 30 pilot plants ranging from 0.5 kW_{th} to 4 MW_{th} that account for more than 5000 hours of operational experience, and stands at a technical readiness level (TRL) of 6. Relevant milestones reached include the demonstration of the possibility to reach 100% of CO_2 capture and 100% of fuel conversion for gaseous fuels, the proof of concept of the CLOU process, and the possibility to obtain high H_2 yields at autothermal conditions including near complete CO_2 capture.

A special mention deserves research on coal combustion as this fuel will play a significant part in the mix of energy production in the near future, being one of the major sources of CO_2 emissions in large power plants. However, the use of renewable feedstocks such as biomass represents an increasing option within the context of the BECCS technologies. In fact, some of the new CL technologies such as CLR, CLG are currently gaining interest in order to produce syngas that could be converted into liquid fuels or hydrogen to be used in the transport sector. It has to be remembered that this represents about one third of the total carbon emissions into the atmosphere.

Up to now, no showstoppers for the implementation of the CL technology at high scale has been detected. The future challenges to its improvement include the development of the pressurized operation, the process scale-up to increase the TRL and maturity of the technology, and the development of long-life oxygen carriers, among others. In any case, CL represents an interesting option within those technologies able to provide solutions to the greenhouse gas effect in the earth.

Acknowledgment

This work was supported by the Spanish Ministry of Economy, Industry and Competitiveness (projects ENE2014-56857-R, ENE2016-77982-R, and ENE2017-89473-R), by the European Regional Development Fund (ERDF), by the CSIC (201780E035), and by the Government of Aragón. T. Mendiara thanks for the "Ramón y Cajal" post-doctoral contract awarded by MINEICO.

References

1. IPCC. Fifth Assessment Synthesis Report.
2. IPCC. IPCC special report on carbon dioxide capture and storage, Cambridge, UK: Cambridge University Press; 2005.
3. IPCC, "Climate Change 2014: Synthesis Report. Contribution of Working Groups I, II and III to the Fifth Assessment Report of the Intergovernmental Panel on Climate Change," Geneva (Switzerland). 2014.
4. Adánez, J.; Abad, A.; García-Labiano, F.; Gayán, P.; de Diego, L.F. (2012) Progress in Chemical-Looping Combustion and Reforming Technologies. *Prog. Energ. Comb. Sci.* 38, 215-282.
5. Adánez, J.; Abad, A.; Mendiara, T.; Gayán, P.; de Diego, L.F. García-Labiano, F. (2018) Chemical Looping Combustion of solid fuels. *Prog. Energ. Comb. Sci.* 65, 6-66.

EELS MEASUREMENT OF GRAPHITIZED ULTRATHIN CARBON FILMS

Noriko Yoshizawa[1*], Yasushi Soneda[1], Masaya Kodama[1]

[1]*Research Institute for Energy Frontier. National Institute of Advanced Industrial Science and Technology (AIST), Tsukuba, Japan*

*Presenting author's e-mail: noriko-yoshizawa@aist.go.jp

Introduction

EELS spectrum of carbon materials are sensitive to the development of graphic structure. C-K edge corresponding to the inner-shell excitation has two characteristic peaks, that is, 1s→π* transition and 1s→ σ* transition. Intensity ratio of these two peaks has been used as a simple parameter to evaluate chemical bonding and consequently development of graphitic structure in the carbon material. On the other hand, incident angle of electron beam to the graphite crystal also affects upon C-K edge feature[1]. It is therefore necessary to consider the overlapping effects of chemical bonding and incident angle of electron beam, especially in case of C-K edge of carbon materials with relatively high crystallinity as graphite. But there seems to be no study of carbon EELS spectrum taking these two factors into account. In this study, we measured EELS spectrum of ultrathin carbon films with parallel/perpendicular orientation of carbon layers, to understand separately the effect of chemical bonding and incident angle of electron beam on C-K edge. EELS spectrum were obtained with "incident beam // carbon layer orientation" or "incident beam ⊥ carbon layer orientation" geometry, and were related to the development of intra-layer structure or stacking structure in the graphitization process of the samples.

Materials and Methods

Poly(benzimidazobenzophenanthroline) (BBL), which is one of the rigid-rod polymers with large molecular plane of ladder-like structure, was used to prepare self-supported ultrathin (10-100 nm order) carbon films by a heat-treatment of spin-coated BBL films up to 3473 K[2]. We have already confirmed mainly by TEM observations that two types of carbon films with different orientation of carbon layers were selectively manufactured from the same starting polymer. That is, one had microtexture in which carbon layers were preferentially oriented perpendicular to the surface of the film (perpendicular-type), and the other had microtexture consisting of stacking structure of carbon layers well developed along the film surface (parallel-type)[3]. Typical cross-section view of TEM for the parallel-type and the perpendicular-type of graphitized ultrathin films were shown in **Figure 1**. Hereinafter, samples were indicated with their type (parallel: //, perpendicular: ⊥) and heat-treatment temperature (HTT). TEM observations of these specimens were carried out at an accelerating voltage of 120 kV. EELS was measured with a spectrometer attached to TEM.

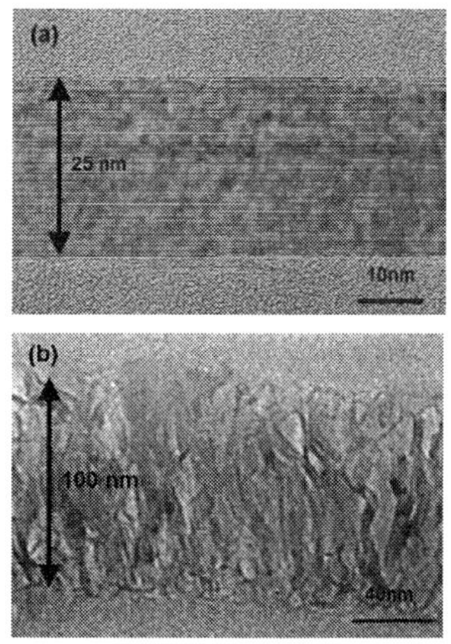

Figure 1. Typical cross-section TEM of graphitized ultrathin carbon films prepare from BBL polymer. (a) parallal- (b) perpendicular- type

Results and Discussion

Figure 2 shows the EELS spectrum of parallel- and perpendicular-type of ultrathin carbon films heat-treated at 3000 °C. These spectrum were recorded in the geometry of electron beam⊥ film surface. In the carbon C-K edge region, characteristic peaks at 285 eV (1s→π* transition) and at 292 eV (1s→ σ* transition) were appeared for both types of films.

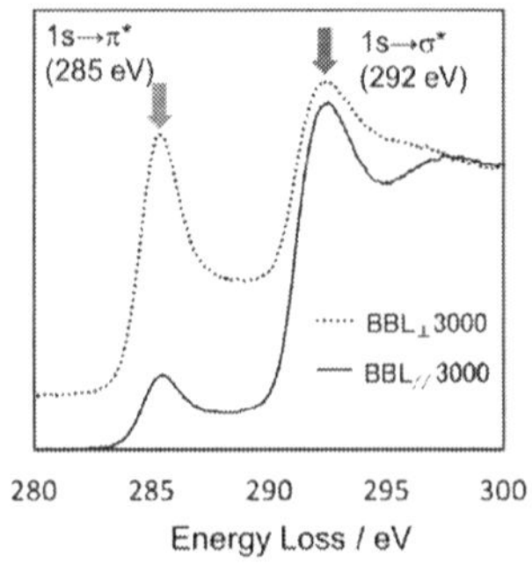

Figure 2. EELS spectrum of parallel- and perpendicular-type of ultrathin carbon films heat-treated at 3000 °C

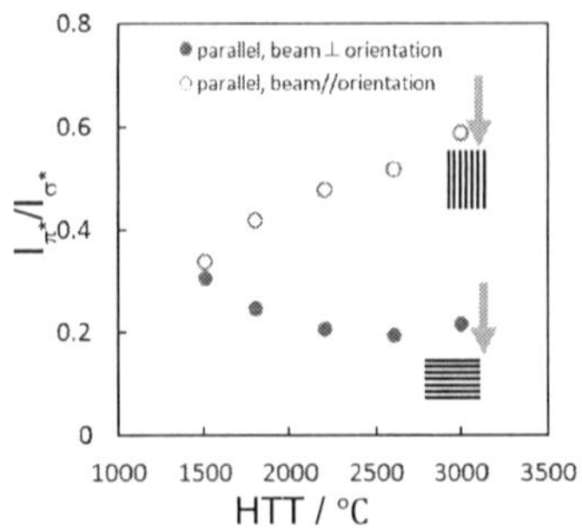

Figure 3. Dependence of I_{π^*}/ I_{σ^*} upon HTT (parallel-type)

Top intensities of these π* transition (I_{π^*}) and σ* transition (I_{σ^*}) peaks were treated here as their peak intensities, and it was clear that their ratio I_{π^*}/ I_{σ^*} was larger in the perpendicular-type film than the parallel type. In order to correlate the ratio I_{π^*}/ I_{σ^*} to the growth of graphitic structure, I_{π^*}/ I_{σ^*} in EELS spectrum taken in the "incident beam // carbon layer orientation" and "incident beam ⊥ carbon layer orientation" was plotted against the heat-treatment temperature (HTT). In **Figure 3**, I_{π^*}/ I_{σ^*} estimated for the parallel type film showed the gradual decrease in beam ⊥ orientation geometry (or increase in beam // orientation geometry) as HTT increased. According to our previous studies on the graphitizing process of these types of films[2], TEM observation of parallel-type film clarified the significant growth of intra-layer structure up to some micrometer along the film surface, as well as development of stacking structure in thickness direction. In beam ⊥ orientation geometry, change in intra-layer structure can be more influential upon the spectrum than that in stacking structure. It means that the increase in I_{σ^*} possibly contributed to the decrease in I_{π^*}/ I_{σ^*} ratio. It is also assumed that the effect of development in stacking structure was more obvious than the growth of intra-layer structure, when EELS was measured in beam // orientation geometry. In this case, the increase in I_{π^*} affected the increase in I_{π^*}/ I_{σ^*}. Discussion upon the I_{π^*}/ I_{σ^*} change was a little complicated for the perpendicular-type film. The perpendicular-type had preferential orientation of carbon layers in thickness direction, but their crystallinity was much lower than the parallel-type. Accordingly, the development of intra-layer as well as stacking structure in their microtexture was not reflected specifically upon their EELS spectrum.

Conclusions

Ultrathin carbon films with parallel or perpendicular orientation of carbon layers were characterized by EELS measurements. Intensity ratio of π* transition to σ* transition peak in C-K edge was used to indicate the characteristics of the EELS spectrum. Considering the results of TEM observations presented in our previous study, the change in I_{π^*}/ I_{σ^*} according to the increase in HTT could be explained by the development of graphite structure.

Acknowledgment

This paper is based on results obtained from a project supported by the New Energy and Industrial Technology Development Organization (NEDO).

References

1. Leapman, R.D, Fejes, P.L., Silcox, J. (1983) Orientation dependence of core edges from anisotropic materials determined by inelastic scattering of fast electrons. *Phys. Rev. B* 28, 2361-2373.
2. Soneda, Y., Odahara, G., Yoshizawa, N. Kodama, M. Synthesis of ultrathin graphite films with controlled preferential orientation of carbon layers. Extended abstracts, CARBON2013.
3. Yoshizawa, N., Soneda, Y., Kodama, M. (2013) TEM study of perpendicular/parallel orientation of aromatic layers in self-supported ultrathin carbon films from BBL polymer. Extended abstracts, CARBON2013.

FROM NANO- TO MACRO-SCALE: HYBRID MAGNETIC CARBON NANOCOMPOSITES AS A TOOL FOR CATALYTIC WET PEROXIDE OXIDATION

Rui S. Ribeiro[1, 2*], Adrián M.T. Silva[2], Joaquim L. Faria[2], Helder T. Gomes[1]

[1] *Laboratory of Separation and Reaction Engineering – Laboratory of Catalysis and Materials (LSRE-LCM), Instituto Politécnico de Bragança, 5300-253 Bragança, Portugal*
[2] *Laboratory of Separation and Reaction Engineering – Laboratory of Catalysis and Materials (LSRE-LCM), Faculdade de Engenharia, Universidade do Porto, 4200-465 Porto, Portugal*

* rui.ribeiro@ipb.pt

Introduction

There is a need for technological innovation in the water sector worldwide, as a result of the growing demand for water supplies – meeting increasingly restricted quality criteria, while coping with the increasing scarcity of clean water sources[1]. Under this context, the use of treated wastewater as an alternative water source has emerged as a topic of high priority[2]. However, meeting current quality requirements for wastewater reuse is a great challenge, in which nanotechnology holds a great potential[1].

Catalytic wet peroxide oxidation (CWPO) is a promising water/wastewater treatment technology; it enables the formation of highly oxidizing hydroxyl radicals ($HO^{\bullet}$) under atmospheric pressure and low to moderate temperatures, when a suitable catalyst is employed for the decomposition of hydrogen peroxide (H_2O_2)[3]. However, further improvement of catalyst design is still required in order to allow the scale-up of the CWPO technology towards real-scale applications.

Bearing this in mind, our work has been focused on the development of highly active and stable hybrid magnetic carbon nanocomposites for CWPO. A detailed catalyst design at the nanoscale, based on the understanding of the surface reactions and interactions involved in the CWPO process, has recently allowed us to move forward towards the treatment of a real industrial wastewater with high pollutant load – collected from a mechanical biological treatment (MBT) plant for municipal solid waste processing. This communication reports the findings obtained in the last four years by our research group in this quest. A particular emphasis is given to the synergistic effects arising from the combination of iron-based catalysts with the easily tuned properties of carbon-based materials.

Results and Discussion

A hybrid magnetic graphitic nanocomposite – composed by a magnetite core and a graphitic shell, was synthesized by hierarchical co-assembly of magnetite nanoparticles and carbon precursors. Following its application in CWPO, it was concluded about the role of the carbon shell. Specifically, the encapsulation of magnetite nanoparticles within carbon frameworks (i) enhances the catalytic activity in CWPO when compared to bare magnetite, while (ii) strongly limiting the leaching of iron species to the treated water. The first effect was ascribed to the increased adsorptive interactions between the carbon phase and the pollutant molecules, which enable a more intimate contact between the generated $HO^{\bullet}$ radicals and the pollutant. The lower metal leaching obtained with the nanocomposite catalyst was ascribed to the confinement effect promoted by the carbon shell. As a result of these positive effects, very high pollutant mass removals were obtained with a rather high efficiency of H_2O_2 consumption, and the composite catalyst was active for operating pH in the range 3 – 6[4].

Seeking for catalyst optimization, the individual effects of the metal species employed in the nanostructured composites have been studied. It was found that the simultaneous incorporation of cobalt and iron into carbon matrices (iii) increases the rate of HO• radicals formation, and (iv) promotes a more efficient reduction of Fe^{3+} to Fe^{2+}, thus limiting iron leaching[5].

Based on the findings reported above, it was possible to design a high performance hybrid magnetic graphitic nanocomposite ($CoFe_2O_4$/MGNC) – composed by a cobalt ferrite core and a graphitic shell. In this way, the positive effects described in (i) to (iv) were combined in the same nanocomposite material. It was found that the application of this new generation catalyst enables the treatment of waste waters with high pollutant loads, such as that from the MBT plant considered in this study (cf. **Figure 1**). The biodegradability of the wastewater was enhanced during the treatment performed at pH 6, regardless of its high organic and inorganic content; disinfection was achieved and the treated water revealed no toxicity against selected bacteria. The high stability of $CoFe_2O_4$/MGNC for CWPO was demonstrated in a series of five CWPO reaction/magnetic separation sequential experiments in the same vessel[6].

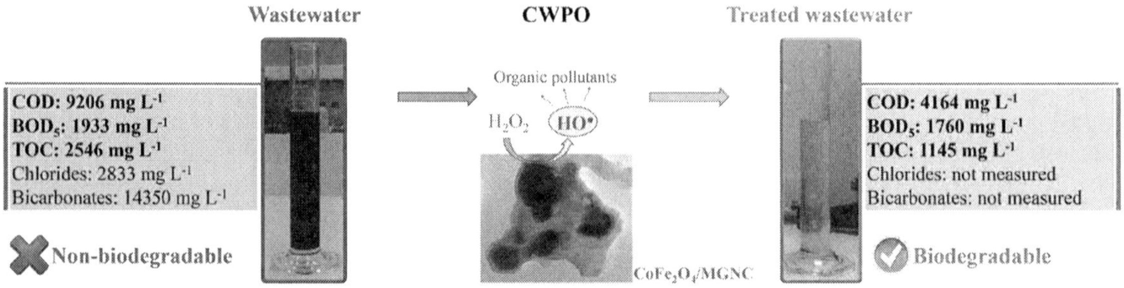

Figure 1. Representation of the CWPO treatment performed in the presence of $CoFe_2O_4$/MGNC.

Conclusions

The proper understanding of the catalyst properties at the nanoscale is fundamental in order to design materials with potential to be an effective tool for real-scale CWPO applications.

Acknowledgment

This work was supported by: Project POCI-01-0145-FEDER-006984 – Associate Laboratory LSRE-LCM funded by FEDER through COMPETE2020 – Programa Operacional Competitividade e Internacionalização (POCI) – and by national funds through FCT – Fundação para a Ciência e a Tecnologia. R.S. Ribeiro acknowledges the FCT individual Ph.D. grant SFRH/BD/94177/2013, with financing from FCT and the European Social Fund (through POPH and QREN).

References

1. Qu, X., Alvarez, P.J.J., Li, Q. (2013). Applications of nanotechnology in water and wastewater treatment. *Water Research* 47, 3931-3946.

2. Fatta-Kassinos, D., Manaia, C., Berendonk, T.U., Cytryn, E., Bayona, J., Chefetz, B., Slobodnik, J., Kreuzinger, N., Rizzo, L., Malato, S., Lundy, L., Ledin, A. (2015). COST Action ES1403: New and Emerging challenges and opportunities in wastewater REUSe (NEREUS). *Environmental Science and Pollution Research* 22, 7183-7186.

3. Ribeiro, R.S., Silva, A.M.T., Figueiredo, J.L., Faria, J.L., Gomes, H.T. (2016). Catalytic wet peroxide oxidation: a route towards the application of hybrid magnetic carbon nanocomposites for the degradation of organic pollutants. A review. *Applied Catalysis B: Environmental* 187, 428-460.

4. Ribeiro, R.S., Silva, A.M.T., Tavares, P.B., Figueiredo, J.L., Faria, J.L., Gomes, H.T. (2017). Hybrid magnetic graphitic nanocomposites for catalytic wet peroxide oxidation applications. *Catalysis Today* 280, 184-191.

5. Ribeiro, R.S., Silva, A.M.T., Figueiredo, J.L., Faria, J.L., Gomes, H.T. (2017). The role of cobalt in bimetallic iron-cobalt magnetic carbon xerogels developed for catalytic wet peroxide oxidation. *Catalysis Today* 296, 66-75.

6. Ribeiro, R.S., Rodrigues, R.O., Silva, A.M.T., Tavares, P.B., Carvalho, A.M.C., Figueiredo, J.L., Faria, J.L., Gomes, H.T. (2017). Hybrid magnetic graphitic nanocomposites towards catalytic wet peroxide oxidation of the liquid effluent from a mechanical biological treatment plant for municipal solid waste. *Applied Catalysis B: Environmental* 219, 645-657.

TiO_2-CARBON HYBRID PHOTOCATALYSTS FOR THE ELIMINATION OF POLLUTANTS IN LIQUID PHASE

Ana Amorós-Pérez[1*], Laura Cano-Casanova[1], Mohammed Ouzzine[1], María Ángeles Lillo-Ródenas[1], M. Carmen Román-Martínez[1], Nicolas Keller[2]

[1]*MCMA Group, Department of Inorganic Chemistry and Materials Institute. Faculty of Science. University of Alicante, Alicante, E-03080, Spain.*
[2]*Institut de Chimie et Procédés pour l'Énergie, l'Environnement et la Santé (ICPEES), University of Strasbourg and CNRS, 67087, France*

*ana.amoros@ua.es

Introduction

The concerns about depletion of non-renewable resources and the increasing contamination of water have contributed to the development of new methods for both energy production and pollutants removal. Photocatalytic reactions stand out for this purpose, being TiO_2 one of the most used and investigated photocatalysts. Combination of TiO_2 with activated carbon (AC) can be beneficial[1,2] because carbon can increase the photocatalyst adsorption capacity and/or decrease the recombination rate of the electron/hole photogenerated pairs[2], among other reasons. Therefore, the present study addresses the preparation of TiO_2-AC hybrid photocatalysts using sucrose as carbon precursor. The behaviour of the prepared photocatalysts is studied in two reactions in aqueous solution, of environmental interest: 1) oxidation of acetic acid (HAc), a frequent residue in different liquid effluents, to produce biogas and, 2) degradation of diuron, an herbicide commonly used in agriculture and responsible for water pollution.

Materials and Methods

An AC with spherical morphology (**Figure 1**) has been prepared from sucrose by hydrothermal synthesis followed by activation with CO_2 (80 mL/min, 5 °C/min, 800 °C, 10 h). TiO_2-AC materials have been prepared *via* sol-gel using titanium tetraisopropoxide (TTIP) and adding the prepared AC during the synthesis to obtained samples with 0, 0.5, 1, 10 wt. % AC (nominal contents). The catalysts have been submitted to a post-synthesis heat treatment (air, 350 °C, 2 h, 10 °C/min). TiO_2 P25 (Evonik) has been used as reference.

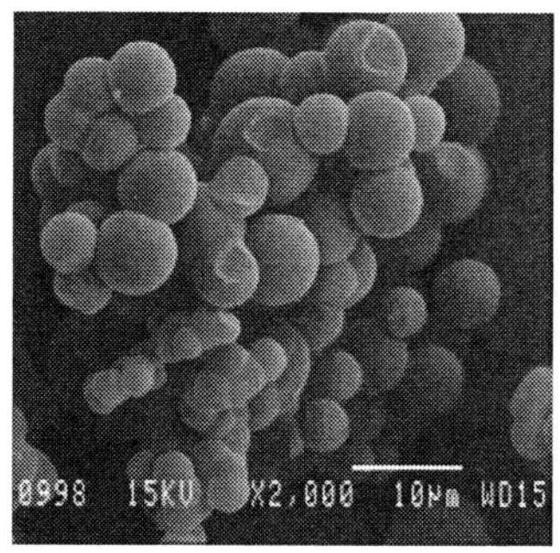

Figure 1. SEM images of the synthesized AC.

The prepared photocatalysts have been characterized by N_2 adsorption, XRD, TG analysis and SEM, and tested in the two reactions mentioned above in the following conditions:

Acetic acid oxidation: 350 mL of 1 M acetic acid, 0.35 g catalyst, room temperature, 12 h, 365 nm Hg lamp. Reaction products, mainly CH_4 and CO_2, have been quantified by mass spectrometry.

Oxidative degradation of diuron: 100 mL of a 10 mg/L diuron solution, 0.1 g catalyst, solar chamber (340 nm lamp). Diuron concentration has been determined by UV-vis spectroscopy. As diuron degradation is a pseudo first-order reaction[3], the experimental results have been fitted to a linear equation to obtain the kinetic constant.

Results and Discussion

Table 1 summarizes the characterization results. The surface area of the prepared materials tends to increase with the carbon content and is larger than that of P25. All the samples only contain anatase as crystalline phase, with crystal size (A_{Size}) around 8-9 nm. Comparing with P25, the percentage of crystalline anatase (A_{Cryst}) in the prepared samples is a bit higher and the crystal size is clearly lower.

Figure 2 compiles the obtained photocatalytic activity data: mmoles of CH_4 and CO_2 produced in the acetic acid oxidation in 12 h and kinetic constant for diuron oxidation.

Regarding acetic acid oxidation, it can be observed that, in general, the prepared samples are more active than P25: the synthetized photocatalysts produce more CH_4 and the CH_4/CO_2 ratio is higher than in the case of the P25 commercial titania. In general, the presence of AC seems to improve the catalysts efficiency respect to that of pure TiO_2. Among the TiO_2-AC catalysts, TiO_2-AC**1** displays the highest activity. Likely, a higher AC loading might interfere with the interaction between the TiO_2 and the light.

Table 1. Textural and crystalline properties and of the prepared materials and of P25.

Sample	S_{BET} (m^2/g)	$V_{DR}(N_2)$ (cm^3/g)	A_{Size}[a] (nm)	A_{Cryst}[b] (%)
AC	491	0.25	--	--
P25	60	0.02	22	70
TiO_2	144	0.05	8	76
TiO_2-AC**0.5**	154	0.06	9	74
TiO_2-AC**1**	151	0.06	8	74
TiO_2-AC**10**	173	0.07	8	74

[a]*A_{Size}: anatase crystal size*
[b]*A_{Cryst}: percentage of crystalline anatase*

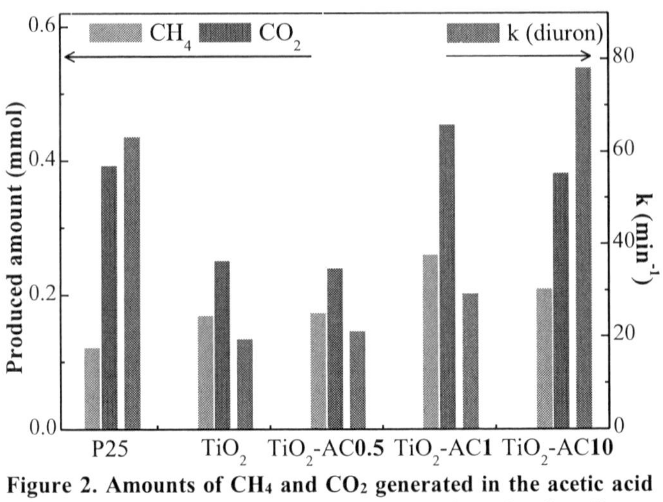

Figure 2. Amounts of CH_4 and CO_2 generated in the acetic acid oxidation (12 h) and kinetic constant values in the diuron oxidation.

In diuron degradation, the catalysts activity follows the order: TiO_2-AC**10** > P25 > TiO_2-AC**1** > TiO_2-AC**0.5** > TiO_2. The carbon-containing catalysts are more active than the prepared pure TiO_2. However, with the exception of sample TiO_2-AC**10**, they are less active than P25. The high activity of catalyst TiO_2-AC**10** is likely related with the adsorption of diuron in the AC.

Conclusions

The prepared TiO_2-AC photocatalysts are more active in the two investigated reactions than pure titania prepared in a similar way. In the acetic acid oxidation, the best results have been obtained with the catalyst containing 1 wt.% AC, while in the degradation of diuron is the catalyst with the highest carbon loading (10 wt.%) the one leading to the highest diuron removal of the catalyst (likely by a combination of adsorption and oxidation processes).

Acknowledgment

The authors thank projects of references CTQ2015-66080-R (MINECO/FEDER), PROMETEOII/2014/010 (GV/Feder) and VIGROB-136 (University of Alicante) for financial support.

References

1. Ouzzine, M., Romero-Anaya, A.J., Lillo-Ródenas, M.A., Linares-Solano, A. (2014). Spherical activated carbon as an enhanced support for TiO_2/AC photocatalysts. *Carbon* 67, 104-118.
2. Kavitha, R., Devi, L.G. (2014). Synergistic effect between carbon dopant in titania lattice and surface carbonaceous species for enhancing the visible light photocatalysis. *J. Environ. Chem. Eng.* 2, 857-867.
3. Kouamé, N.A., Robert, D., Keller, V., Keller, N., Pham, C., Nguyen, P. (2011). Preliminary study of the use of β-SiC foam as a photocatalytic support for water treatment. *Catal. Today* 161, 3-7.

Multiphoton-assisted surface plasmon-coupled emission in DNA-assembled nanodiamonds

Morteza Aramesh[1*], Raphael Tiefenauer[1], Volker Gatterdam[1], Lukas Schmidheini[1], Takumi Sannomiya[2]

[1] *Laboratory of Biosensors and Bioelectronics, ETH Zürich, 8092 Zürich, Switzerland*
[2] *School of Materials & Chem.Tech., Dept. Materials Science and Engineering, Tokyo Institute of Technology, 226-8503 Yokohama, Japan*
**aramesh@biomed.ee.ethz.ch*

Introduction

Nanodiamond particles (NDs) are emerging as new promising agents for various biomedical and/or quantum information applications.[1–3] NDs exhibit several unique properties, such as superior chemical stability, non-toxicity and tuneable surface functionality. Furthermore, atomic defects can be engineered within the crystal structure of NDs with high precision, allowing generation of stable and bright color centers which are able to emit single photons at room temperature. These unique combinations of properties open a wide range of opportunities for the use of NDs in drug delivery, intra-cellular imaging, lasing and sensing.[4] Moreover, coupling of NDs to plasmonic structures has shown a great promise for some advanced applications such as quantum communication and sensing.[5–7] However, scalable assembly of such hybrid structures faces several technical challenges.[8]

Light emission inherent in nanodiamonds is due to the atomic defects in the crystal structure, such as nitrogen vacancies (NV centers). NV centers exhibit photoluminescence emission within the red to near-infrared range of the spectrum (650-760 nm). The photoluminescence is conventionally excited by a green laser light. More recently, nanoscopic excitation of these centers has been demonstrated via coupling to a nanowire waveguide.[9] However, the possibility of excitation of NV centers with localized surface plasmons (LSPs) has not been demonstrated.

Here we present a new physical system in which nanodiamonds are optically coupled to a plasmonic component, i.e. gold nanoparticles (GNPs), via DNA-based assembly. We show that the energy of the localized surface plasmons in GNPs can be transferred to the ND particles in the close proximity. A multi-photon process was implemented to generate localized surface plasmons in GNPs. The near-field resonance of the LSPs serves as an energy source for excitation of optical centers in NDs.

Results and Discussion

The principle of DNA-induced nanoassembly is shown in **Figure 1**. DNA-functionalized particles were bound together via hybridization with a complementary DNA. The complementary DNA is composed of two parts, in which one half of the sequence is complementary to the GNP tags and the other half to the ND tags. Upon DNA hybridization, the conjugated ND-GNP pairs form in the solution. The conjugated assemblies then can be spin coated on a glass substrate for optical analysis.

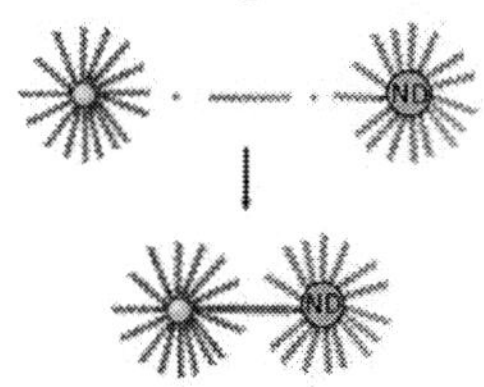

Figure 1. Principles of the DNA-induced nanoassembly. DNA-functionalized nanoparticles form conjugates via hybridization by a complementary DNA that contains complementary sequences to both GNP and ND tags.

GNPs exhibit plasmonic resonances when illuminated with visible light. The resonance band of the LSPs in GNPs are size-dependent.[10] For example, the GNPs used in this study (50 nm in size), show a LSP resonance at 520 nm, which is very close to the wavelength required for NDs excitation. We investigated suitability of multi-photon (MP) microscopy for specific excitation of GNPs. In the first step, we studied the optical properties of each component of the hybrid structure individually.

Figure 2 shows the emission of GNPs and NDs under MP illumination at various wavelengths. While no fluorescent emission was observed in NDs at the investigated wavelengths, GNPs on the other hand, showed wavelength-dependent emission. Interestingly, at λ=1000 nm excitation, GNPs exhibited a strong emission centered at 520 nm, which perfectly matches with the LSP resonance of the particles. The emission power was quadratically proportional to the excitation power, suggesting that the emission is due to the plasmonic excitation of the particles via two-photon absorption.

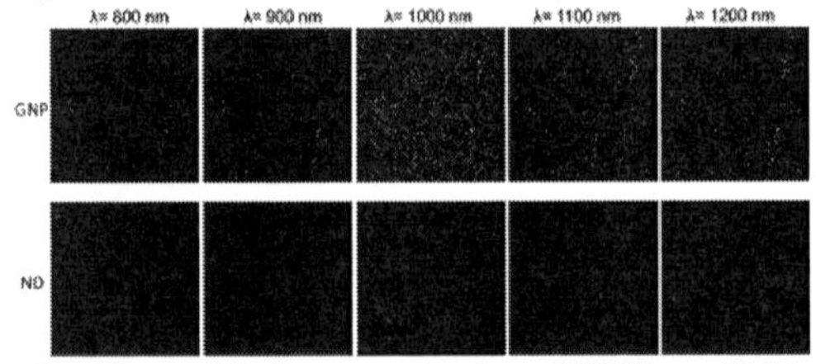

Figure 2. Optical images obtained by multi-photon excitation of GNPs and NDs at different wavelengths.

In the next step, we examined the optical coupling between the hybrid GNP-ND pairs under multi-photon illumination (**Figure 3**). Interestingly, the collected spectrum obtained from the hybrid particles consisted of two peaks at 520 and ~600-700 nm, which are correlated with SPR of GNPs and photoluminescence of NDs, respectively. Based on this observation, we suggest that the excitation of NDs is due to the near-field excitation by the neighboring GNPs.

Figure 3. Optical images of the coupled GNP-ND system under multi-photon excitation. The plasmonic emission from GNPs contributes in excitation of neighboring NDs.

Conclusions

In summary, we introduced a new physical system in which GNPs and NDs form conjugates via DNA hybridization. The conjugated particles exhibit interesting optical properties. Particularly, we demonstrated that the fluorescent emission in NDs can be coupled to the surface plasmon resonances of GNPs generated via a multi-photon process. The introduced system can be implemented for biosensing in the future.

References

[1] J. Gong, N. Steinsultz, M. Ouyang, Nanodiamond-based nanostructures for coupling nitrogen-vacancy centres to metal nanoparticles and semiconductor quantum dots, Nat. Commun. 7 (2016) 1–10.

[2] W.W. Hsiao, Y.Y. Hui, P.C. Tsai, H.C. Chang, Fluorescent Nanodiamond: A Versatile Tool for Long-Term Cell Tracking, Super-Resolution Imaging, and Nanoscale Temperature Sensing, Acc. Chem. Res. 49 (2016) 400–407.

[3] F.N. Gür, C.P.T. McPolin, S. Raza, M. Mayer, D.J. Roth, A.M. Steiner, M. Löffler, A. Fery, M.L. Brongerma, A. V. Zayats, T.A.F. König, T.L. Schmidt, Self-assembled plasmonic waveguides for excitation of fluorescent nanodiamonds, (2017). http://arxiv.org/abs/1712.09141.

[4] X.Q. Zhang, M. Chen, R. Lam, X. Xu, E. Osawa, D. Ho, Polymer-functionalized nanodiamond platforms as vehicles for gene delivery, ACS Nano. 3 (2009) 2609–2616.

[5] A. Huck, S. Kumar, A. Shakoor, U.L. Andersen, Controlled coupling of a single nitrogen-vacancy center to a silver nanowire, Phys. Rev. Lett. 106 (2011) 2–5.

[6] M. Aramesh, J. Cervenka, A. Roberts, A. Djalalian-Assl, R. Rajasekharan, J. Fang, K. Ostrikov, S. Prawer, Coupling of a single-photon emitter in nanodiamond to surface plasmons of a nanochannel-enclosed silver nanowire., Opt. Express. 22 (2014) 15530–41.

[7] Z. Al-Baiaty, B.P. Cumming, X. Gan, M. Gu, Detection of the ODMR signal of a nitrogen vacancy centre in nanodiamond in propagating surface plasmons, J. Opt. (United Kingdom). 20 (2018).

[8] O. Benson, Assembly of hybrid photonic architectures from nanophotonic constituents, Nature. 480 (2011) 193–199.

[9] J. De Torres, P. Ferrand, G. Colas Des Francs, J. Wenger, Coupling Emitters and Silver Nanowires to Achieve Long-Range Plasmon-Mediated Fluorescence Energy Transfer, ACS Nano. 10 (2016) 3968–3976.

[10] J. Zheng, C. Zhou, M. Yu, J. Liu, Different sized luminescent gold nanoparticles, Nanoscale. 4 (2012) 4073.

MATERIAL AND PROCESSING VARIABLES AFFECTING THE PROPERTIES OF CARBON BONDED CARBON FIBER INSULATION

Nidia C. Gallego[1*], Glenn R. Romanoski[1], Kyle A. Lach[2], Cristian I. Contescu[1], Shiba Adhikari[1], Ashli M. Clark[1], and George B. Ulrich[1]

[1] *Materials Science and Technology Division, Oak Ridge National Laboratory, Oak Ridge TN, USA*
[2]*Department of Chemical Engineering, University of Dayton, Dayton, OH, USA*

*Presenting author's e-mail: gallegonc@ornl.gov

Introduction

The requirement for containment and thermal protection of the isotopic fuel in radioisotope power systems led to a design and material selection for the General Purpose Heat Source (GPHS) that relied principally on carbon materials. A unique Carbon Bonded Carbon Fiber (CBCF) insulation was developed to provide thermal protection during the extremes of unlikely reentry scenarios. The two principal constituents of CBCF are chopped aerospace-grade rayon and a powdered phenolic resin. The principal physical properties that define the performance of CBCF as an insulator in the GPHS include: density, thermal conductivity and compressive strength.

The key steps in CBCF production include[1,2]: chopping of continuous rayon tow, carbonization of the chopped rayon fibers, prewetting of fibers, fiber–resin–water slurry preparation, vacuum molding of cylindrical and plate fiber-resin preforms and, drying, curing and carbonization of these preforms (see Figure 1). Although the raw materials and processing steps have remained nominally constant over two decades, some variability in final densities and thermal properties has motivated a detailed investigation of starting materials and processing. We will report our findings regarding the effects of fiber length distribution for the chopped rayon and chopped and carbonized fibers at various stages in processing, as well as some insight into the melt, flow and cure behavior of the phenolic resin obtained from the implementation of several advanced techniques.

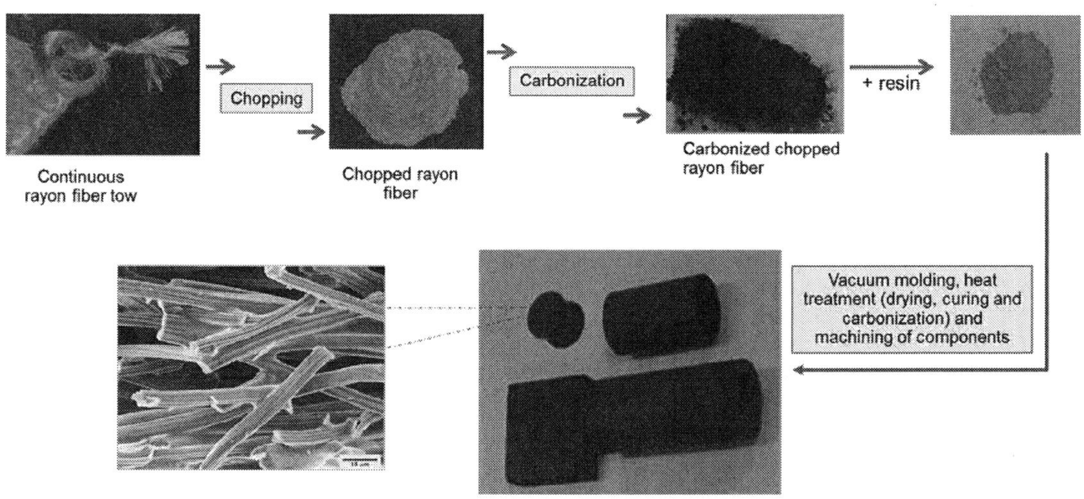

Figure 1. Schematic of the production of CBCF material

Materials and Methods

The principal constituents of CBCF insulators are carbonized rayon fiber and phenolic resin. The rayon fibers are chopped and carbonized at high temperature under a protective atmosphere. The first step in preparation of the carbon fiber/phenolic resin/water slurry involves wetting of the carbon fibers. Since the chopped and carbonized fiber generally resists wetting, a considerable amount of mechanical energy is employed, including mechanical mixing using a stirrer and ultrasonic horn, followed by ball milling using yttria-stabilized zirconia milling media. Finally, the wetted carbon fibers are added to the large slurry tank with the pre-wetted phenolic resin powder in preparation for vacuum slurry molding. After molding, the parts are dried and cured in a convection oven following a heat treatment cycle that last over 14 hours. A detailed investigation of the effect that the high mechanical energy imparted during the fiber wetting process on the length distribution of the carbonized rayon fibers was conducted utilizing an automated image analysis system. In addition, the macroscopic flow behavior of the resin during drying and curing is being studied using an industry standard resin slump test and a hot stage microscope.

Results and Discussion

Figure 2 shows the fiber length distribution of the carbonized rayon fibers before the wetting process and after the mechanical mixing and ball milling processes. As observed, the fiber length distribution is changed significantly during these two steps used for wetting of the carbon fibers prior to preparing the slurry. We are evaluating the effect of various stirrer geometries, positions of the ultrasonic horn and milling time on the fiber length and distribution. Additional, several CBCF samples are being prepared with fibers processed in different ways to find a correlation between these variables and the final CBCF density.

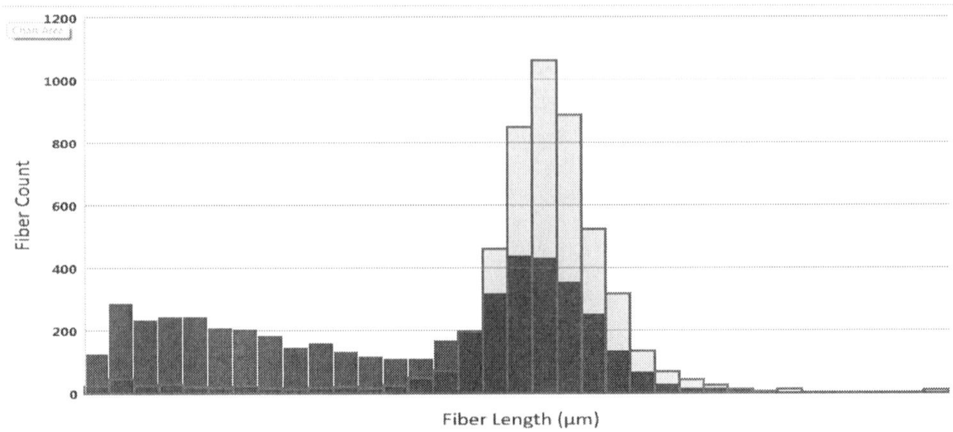

Figure 2. Carbonized rayon fiber length distributions before wetting (red) and after wetting process (blue)

Conclusions

Detailed analysis of rayon fiber length distributions throughout CBCF processing has identified unintended fiber fracturing to be a controlling factor in the production process that significantly affects the density and properties of the final insulators.

Acknowledgment

This work was sponsored by the United States Department of Energy Office of Facilities Management (NE-31). Primary funding was provided by NASA Science Mission Directorate.

References

1. A. Schock, "Design Evolution and Verification of the General Purpose Heat Source," Intersociety Energy Conversion Engineering Conference, Proceedings, pp. 1032-1042, 1980
2. G. C. Wei and J. M. Robbins, "Carbon-Bonded Carbon Fiber Insulation for Radioisotope Space Power Systems", Ceramic Bulletin, 64 (5), 1985

ROSIN-MEDIATED ULTRACLEAN TRANSFER OF GRAPHENE AND RELATED TWO-DIMENSIONAL MATERIALS

Jinhong Du[1*], Dingdong Zhang, Wencai Ren, Hui-Ming Cheng

Institute of Metal Research, Chinese Academy of Sciences, 72 Wenhua Road, Shenyang 110016, China

* jhdu@imr.ac.cn

Introduction

The large polymer particle residue generated during the transfer process of graphene and related two-dimensional (2D) materials grown by chemical vapor deposition (CVD) is a critical issue that limits their use in large-area thin-film devices such as thin film transistor, organic light-emitting diodes (OLEDs) and organic photovoltaic cells. [1-3] This is because that the commonly used macromolecular polymer support materials during transfer process have low solubility, and strong interaction with graphene and 2D related materials, therefore, are difficult to be removed after the transfer. The resulted polymer residues can generate a large surface roughness, which often results in a high leakage current and even causes a short circuit of devices. Thus, it is highly desired to find the new support materials that can be easily removed from the surface of graphene and 2D related materials after transfer.

In this work, we report a natural organic small molecular rosin as a support layer, whose weak interaction with graphene, good solubility and sufficient strength enable ultraclean and damage-free transfer. Such ultraclean graphene has produced the four-inch monolithic flexible graphene-based OLED with uniform light emitting and a high brightness of about 10,000 cd m^{-4}. Furthermore, CVD-grown 2D WSe_2 and WS_2 from Au to target substrate are achieved with the assistance of rosin support layer. This rosin mediated transfer method provides a universal approach for the ultraclean and damage-free transfer of graphene and related 2D materials grown by CVD on metals.

Materials and Methods

Monolayer graphene film was grown on copper foils. Then, the graphene films were transferred to polyethylene terephthalate (PET) substrate by spin-coating a thin layer of rosin (average Mw ca. 302 by gel permeation chromatogram, Alfa-Aesar CAS no. 8050-09-7, dissolved in ethyl lactate with a concentration of 50 wt.%) at 500 r.p.m. for 10 s and then at 1,200 r.p.m. for 60 s as support layer. Finally, the rosin layer was dissolved by warm acetone (Analytical reagent, 99%, 50 °C) for three times, rinsed with isopropanol and dried by a high purity N_2 flow.[4] To prove the universality of rosin mediated transfer method, WSe_2 and WS_2 grown on Au by CVD were also transferred with the assistance of rosin support layer.

Results and Discussion

We used scanning electron microscopy (SEM) and atomic force microscopy (AFM) to characterize the surface of rosin-transferred graphene films. It was found that the rosin-transferred graphene films are ultraclean. There is nearly no rosin residue can be observed by SEM (Figure 1a) and AFM (Figure 1b). We then used AFM to characterize the surface roughness of the transferred graphene films. Only few small particles are occasionally observed on the rosin-transferred sample, with a small root mean square (RMS) roughness of 0.66 nm and a the maximum height (R_{max}) of the large residue particles of about 15nm, showing the advantage of rosin as a support layer for the clean transfer of graphene. A rosin-transferred monolayer graphene film of size 10×10 cm^2 was divided into 100 equal areas for electrical property measurements. All these areas show a very uniform sheet resistance of 560 Ω/□

with a standard deviation of about 1%, indicating the good uniformity of the transferred graphene film in large area.

To further show the advantages of our rosin transfer method, we tried to fabricate a four-inch monolithic flexible green OLED with rosin-transferred five-layer graphene films as anode. The whole device can be lighted at about 5V. The luminescence is very uniform over the whole four-inch lighting area. At an applied voltage of 16 V, the OLED device shows a brightness in excess of 10,000 cd m^{-2}, which already satisfies the requirements of lighting sources and displays and is even better than some commercial-off-the-shelf OLED panels.

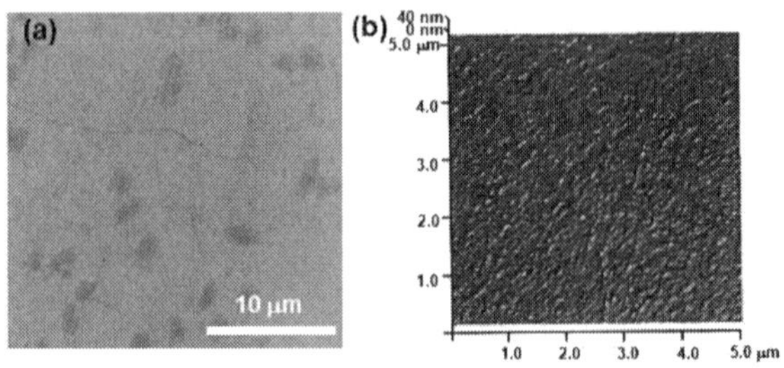

Figure 1. The transferred graphene film using rosin as support layer. (a) SEM and (b) AFM images of graphene transferred on SiO_2/Si substrate

To prove the universality of rosin mediated transfer method, WSe_2 and WS_2 grown on Au by CVD were also transferred with the assistance of rosin support layer. The transferred triangle WSe_2 and WS_2 domain remains intact and is very clean, enabling to a uniform thickness and average surface roughness.

Conclusions

Ultraclean transfer of large-area CVD-grown graphene films has been achieved using the small organic molecule rosin as a support layer. The transferred graphene films have a very low surface roughness with a maximum height of residue particles up to 15 nm and an extremely uniform sheet resistance of 560/□with about 1% deviation over a large area. Such clean graphene enabled to production of the 4-inch monolithic flexible graphene-based OLED exhibiting uniform light emission and a high brightness of about 10,000 cd m^{-2}. Furthermore, CVD-grown 2D WSe_2 and WS_2 from Au to target substrate is achieved with the assistance of rosin support layer, proving the universality of the rosin-based transfer method, which paves the way for electronic and optoelectronic applications, in particular, large-area thin-film devices of graphene and related 2D materials.

Acknowledgment

This work was supported by the National Science Foundation of China (Nos. 51572265, 51861135201, 51521091), the Ministry of Science and Technology of China (No. 2016YFA0200101), and the Chinese Academy of Sciences (No. KGZD-EW-303-3).

References

1. Du, J. H., Jin, H., Zhang, Z. K., Zhang, D. D., Jia, S., Ma, L. P., Ren, W. C., Cheng, H. M., Burn, P. (2017) Efficient organic photovoltaic cells on a single layer graphene transparent conductive electrode using MoOx as an interfacial layer. *Nanoscale* 9, 251-257

2. Jia, S., Sun, H. D., Du, J. H., Zhang, Z. K., Zhang, D. D., Ma, L. P., Chen, J. S., Ma, D. G., Cheng, H. M., Ren, W. C. (2016) Graphene oxide/graphene vertical heterostructure electrodes for highly efficient and flexible organic light emitting diodes. *Nanoscale* 8(20), 10714-10723.

3. Du, J. H., Pei, S. F., Ma, L. P., Cheng, H. M. (2014) Carbon nanotube- and graphene-based transparent conductive films for optoelectronic devices. *Adv. Mater.* 26 (13), 1958-1991.

4. Zhang, Z. K., Du, J. H., Zhang, D. D., Sun, H. D., Yin, L. C., Ma, L. P., Chen, J. S., Ma, D. G., Cheng, H. M., Ren, W. C. (2017) Rosin-enabled ultraclean and damage-free transfer of graphene for large-area flexible organic light-emitting diodes. *Nat. Commun.* 8, 14560.

THE STATE OF WATER ADSORBED IN CARBON NANOPORES

Krisztina László,[1*] Orsolya Czakkel[2], Andrea Domán[1], Erik Geissler[3]

[1] *Physical Chemistry and Materials Science. Budapest University of Technology and Economics, H-15521 Budapest, Hungary*
[2] *Institut Max Von Laue - Paul Langevin, CS 20156, F – 38042 Grenoble Cedex 9, France*
[3]*LIPhy, Université de Grenoble Alpes, CNRS, 38041 Grenoble cedex*

* Presenting author's e-mail: klaszlo@mail.bme.hu

Introduction

Removal of volatile organic vapour from the air is still one of the leading applications of high surface area nanostructured carbons. As such carbons are intrinsically general adsorbents, the relative humidity of air under ambient conditions can compromise the performance of the carbon bed. For this reason, recognition of the role of water is particularly pertinent in activated carbon – vapour phase interactions. The structure of molecular liquids confined inside the nanopores of a high porosity material may differ significantly from that in the bulk liquid.[1] Here we investigate the structure of water vapour adsorbed in two different nanoporous carbons, by means of small angle X-ray and neutron scattering (SAXS and SANS), nitrogen adsorption at 77 K, water vapour adsorption at 293 K, pulsed field gradient NMR, and thermogravimetry.

Materials and Methods

The two samples investigated came from the same batch of high surface area granular carbon (R1 Extra, Norit). The sample designated R1 was as supplied, while sample OX was exposed to concentrated nitric acid for 3 h at room temperature, then thoroughly washed with distilled water and dried. For the SAXS measurements the samples were powdered, placed in capillary tubes and exposed to water vapour at various relative humidity RH, then sealed.[2] For the SANS observations the ground samples were placed in boron-free glass tubes and exposed to D_2O vapour.[3] The NMR observations were made on these carbons exposed to water vapour at RH=45% and 85%. Water and D_2O exposure times were in each case greater then 4 weeks. Experiments were performed on the European Photon and Neutron (EPN) Campus in Grenoble France at the Institut Laue-Langevin (SANS) and the European Synchrotron (ESRF).

Results and Discussion

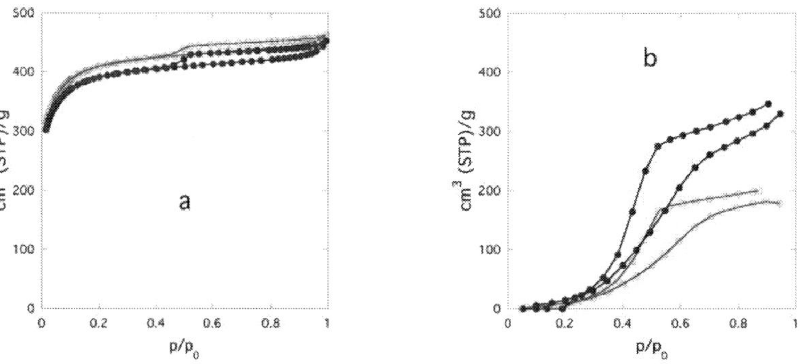

Figure 1 Adsorption/desorption isotherms of the granular high surface area carbon samples. a) nitrogen at 77 K; b) water, 298 K. ○: R1, ●: OX

The nitrogen and water vapour isotherms of the two carbons in **Figure 1 a** and **b** show that in spite of their different surface chemistry the morphology of the two carbons is very similar.

Figure 2a compares the difference in SAXS intensity at high transfer wave number q between the dry sample OX and the same sample exposed to water vapour at RH=85% with that of bulk water. This figure shows that the structure factor $S(q)$ of the adsorbed water differs appreciably from that of bulk water. **Figure 2b** shows that while deuterated toluene contrast matches the SANS signal of carbon, both in the nanopores ($0.2<q<0.6$ Å^{-1}) and in the mesopores ($q<0.2$ Å^{-1}), with water vapour this matching holds only in the nanopores: the steep power-law response at low q comes from hydrophobic surfaces that the water fails to wet.

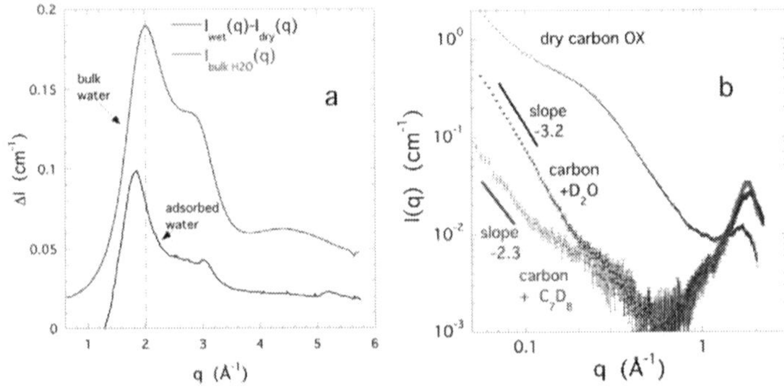

Figure 2. a) Signal of adsorbed water vapour from SAXS compared with that of bulk water; b) SANS signals from dry carbon OX (black), with D2O at RH=85% (blue), and with deuterated toluene vapour (red).

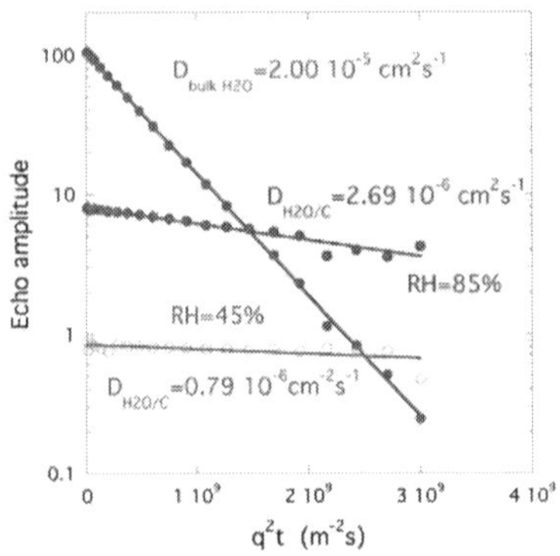

Figure 3 Pulsed field gradient NMR of water adsorbed in sample OX

Pulse field gradient NMR measurements show (**Figure 3**) that the translational diffusion coefficient D of adsorbed H_2O is an order of magnitude slower than in bulk water at RH 85%, and a further factor of 3 slower at RH=45%. The residence time $\tau=1/\{\Delta\omega^2 T_2\}$ of the H_2O molecules increases by a factor 30 and 100 respectively, where T_2 is the nuclear transverse relaxation time, and $\Delta\omega^2$ the dipolar broadening.

Conclusions

Water molecules adsorbed in the nanopores of the activated carbons display liquid like behaviour, but their arrangement is notably different from the bulk state. The translational diffusion coefficient D is severely reduced while the residence time τ of the molecules at adsorption sites greatly enhanced.

Acknowledgments

Access to the French CRG beamline BM02 at the ESRF to the SANS beam lines D11 and D16 at the ILL and to the Bruker mq20 NMR instrument at the ILL is gratefully acknowledged. NMR measurements were done in the Partnership of Soft Condensed Matter laboratories of the EPN campus on an instrument bought by the BMBF grant No. 05K13KT1.The research within the project VEKOP-2.3.2-16-2017-00013 (Managing Authority for Regional Development Programmes, Ministry for National Economy, Hungary) was supported by the European Union and the State of Hungary, co-financed by the European Regional Development Fund.

References

1. Bellissent-Funel M.-C., Sridi-Dorbez R., Bosio L. (1996) X-ray and neutron scattering studies of the structure of water at a hydrophobic surface *J. Chem. Phys.* 104 10023-10029.
2. László, K., Czakkel, O., Dobos, G., Lodewyckx, P., Rochas, C., Geissler, E. (2010) Water vapour adsorption in highly porous carbons as seen by small and wide angle X-ray scattering *Carbon* 48, 1038 –1048
3. László, K, Czakkel, O; Demé, B; Geissler, E (2012). Simultaneous adsorption of toluene and water vapor on a high surface area carbon *Carbon* 50, 4155-4162.

COAL FLY ASH DERIVED NANOCOMPOSITE FOR MERCURY REMOVAL FROM WASTEWATER

Zhandos Tauanov[1], Dhawal Shah[1], Grigorios Itskos[2], Vasileios Inglezakis[1], Lyuba Mikhalovska[3], Sergey Mikhalovsky[3,4]*

[1] *School of Engineering, Nazarbayev University, Astana, Kazakhstan*
[2] *School of Materials Engineering, Purdue University, West Lafayette, IN, USA*
[3] *School of Pharmacy and Biomolecular Sciences, University of Brighton, Brighton, UK*
[4] *ANAMAD Ltd, Brighton, UK*

*sergeymikhalovsky@gmail.com

Introduction

Coal fly ash (CFA) derived materials have become popular with recent advances and expanding range of applications, particularly as a low-cost and easy-processing adsorbent for wastewater treatment and gas purification[1–3]. Among these applications, perpetual interest has been in utilisation of CFA derived composite materials that contain metal and metal oxide nanoparticles for removal of heavy metals from wastewater[4]. We herein focus on exploring the use of locally available CFA and its derivative composite containing Ag nanoparticles (NPs) for removal of mercury (II) from wastewater.

Materials

Representative CFA samples were obtained from Semey power plant in East Kazakhstan, silver nitrate ($AgNO_3$, 99.9%) and sodium hydroxide (NaOH, 99.5%) were supplied by Sigma-Aldrich, sodium borohydride ($NaBH_4$, 99.9%) and mercury (II) chloride ($HgCl_2$, 99.9%) were purchased from Fischer-Scientific. Concentrated and dilute solutions of nitric acid and sodium hydroxide were used for pH adjustments suitable for batch adsorption experiments.

Methods

CFA underwent an alkaline hydrothermal treatment at 110°C, using 3M aqueous NaOH as an activation solution. The incubation period was set at 48 h and a mixing rate of 125 rpm in all experiments. After set period, the mixture was filtered and washed several times until no NaOH was detected. The solid residue was dried at 70°C overnight. The produced zeolite (ZFA) was loaded with Ag NPs as follows: firstly, ZFA was impregnated with silver ions by adding an appropriate volume of 10 mM aqueous $AgNO_3$ solution (1-10 mL per g of ZFA) and soaking for 12 h. Then, the soaked zeolite slurry was dried at 130°C for 3 h and the same soaking procedure was repeated to produce fully ion-exchanged zeolites. These Ag^+-containing zeolites were dried at 130°C for 3 h, filtered and washed with deionised water. After that, Ag^+-ZFA were calcined at 500°C for 3 h followed by 4 h of reduction of silver ions to Ag NPs using $NaBH_4$ at 5:1 ratio to Ag to produce nanocomposites with various Ag NPs loadings (Ag-ZFA). The same set of experiments were done without adding silver nitrate and the resulting material was labelled as R-ZFA.

Results and Discussion

Figure 1 shows the TEM image of Ag-ZFA nanocomposite and adsorption kinetics of Hg(II) by different zeolites. According to the data, the Ag NPs are of spherical shape with 5-40 nm diameter distributed evenly within the zeolite structure. The nanocomposite material reveals a much faster and strong adsorption compared with parent materials which is demonstrated by the respective adsorption kinetics. We hypothesise that the dominating mechanism of Hg adsorption in zeolites is physical and ion exchange adsorption, whereas the nanocomposite consumes mercury via a redox reaction between Ag and Hg(II) forming an amalgam due to close redox potentials of these metals (0.80/0/85V).

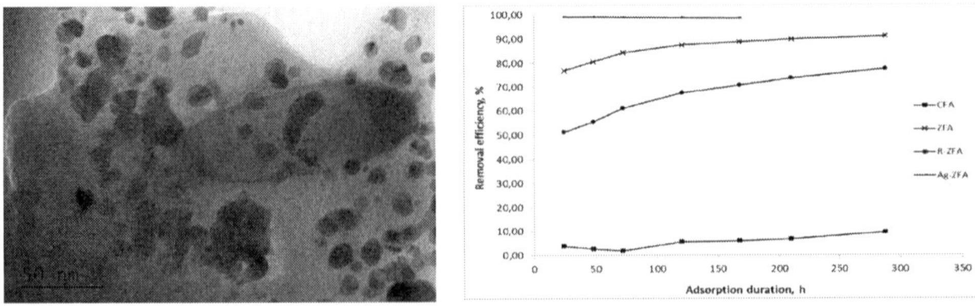

Figure 1. TEM (left) and adsorption kinetics (right) of Hg (II) from water using CFA, zeolite and nanocomposite

The XRD data of the initial zeolite and respective nanocomposite are shown in Figure 2. The results reveal that characteristic peaks of silber in the Ag NP nanocomposite decrease or disappear upon reaction with Hg (II). Similarly, Ag NP peak disappearance upon interaction with mercury was observed elsewhere[5], where authors established the formation of paraschachnerite Ag_3Hg_2 that has an orthorhombic crystal structure. Hg^{2+} reduction and amalgamation were also observed by other authors, who suggested a possible reaction mechanism between Hg^{2+} and Ag NPs. According to this mechanism, an amalgam formation and reduction follow the reactions below:

$$Ag_n + Hg^{2+} \rightarrow Ag_{(n-2)}Hg + 2Ag^+$$

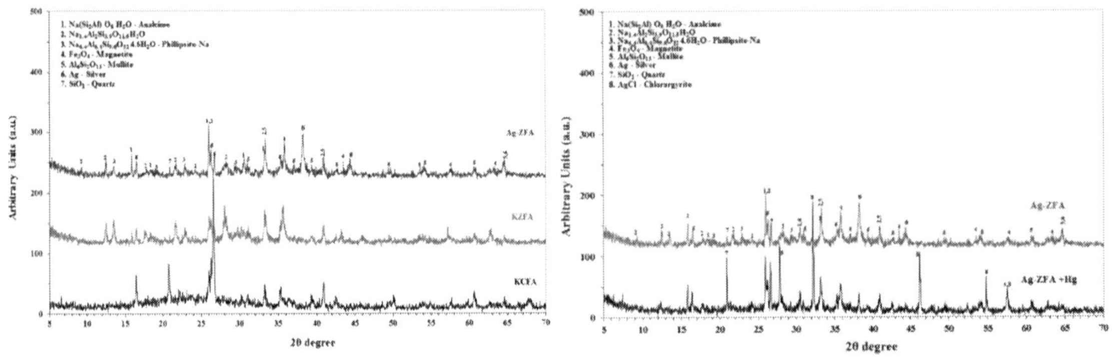

Figure 2. XRD of zeolite and Ag NP nanocomposite before (left) and after (right) Hg removal

Conclusions

Ag NPs impregnated composite derived from CFA was successfully applied for Hg (II) removal from water. The adsorption kinetics of Hg (II) by the nanocomposite is fast and highly effective. The dominating mechanism of Hg uptake in zeolites is via physical and ion exchange adsorption, whereas the Ag NP nanocomposite retains mercury in an amalgam following its reduction to the Hg(0) state.

Acknowledgements

This work is funded by HYPERActive project of Nazarbayev University, the project code SOE2015009, and the Royal Academy of Engineering (UK), project IAPP/1516/13.

References

1. Derkowski, A., Franus, W., Beran, E. & Czímerová, A. (2006). Properties and potential applications of zeolitic materials produced from fly ash using simple method of synthesis. *Powder Technol.* 166, 47–54.
2. Attari, M., Bukhari, S. S., Kazemian, H. & Rohani, S. (2017). A low-cost adsorbent from coal fly ash for mercury removal from industrial wastewater. *J. Environ. Chem. Eng.* 5, 391–399.
3. Jha, V. K., Matsuda, M. & Miyake, M. (2008). Sorption properties of the activated carbon-zeolite composite prepared from coal fly ash for Ni2+, Cu2+, Cd2+ and Pb2+. *J. Hazard. Mater.* 160, 148–153.
4. Xie, J., Wang, Z., Wu, D. & Kong, H. (2014). Synthesis and properties of zeolite/hydrated iron oxide composite from coal fly ash as efficient adsorbent to simultaneously retain cationic and anionic pollutants from water. *Fuel* 116, 71–76.
5. Sumesh, E., Bootharaju, M. S., Anshup & Pradeep, T. (2011). A practical silver nanoparticle-based adsorbent for the removal of Hg2+ from water. *J. Hazard. Mater.* 189, 450–457.

RAPID ADSORPTION OF PRO-INFLAMMATORY CYTOKINES BY GRAPHENE NANOPLATELETS AND THEIR COMPOSITES

Sergey Mikhalovsky[1,2]*, Yishan Zheng[1], Nicholas Pescatore[3], Yury Gogotsi[3], Boris Dyatkin[3], Ganesh Ingavle[1], Vadym Mochalin[4], Tochukwu Ozulumba[1], Susan Sandeman[1]

[1] *School of Pharmacy and Biomolecular Sciences, University of Brighton, Brighton, UK*
[2]*ANAMAD Ltd, Brighton, UK*
[3] *Department of Materials Science and Engineering, Drexel University, Philadelphia, PA, USA*
[4] *Department of Chemistry and Department of Materials Science & Engineering, Missouri University of Science & Technology, Rolla, MO, USA*

* sergeymikhalovsky@gmail.com

Introduction

Sepsis is a complex clinical syndrome that features excessive release of cytokines and other inflammatory mediators that could lead to organ dysfunction. Despite different treatment and management options, sepsis remains associated with high morbidity and mortality rates. This has prompted intensive research into alternative therapeutic approaches such as targeted removal of sepsis related molecules using extracorporeal hemoperfusion. In this study, we explore the use of graphene nanoplatelets (GNP) as a low cost alternative haemosorbent for rapid removal of a broad spectrum of pro-inflammatory cytokine markers.

Materials and Methods

The GNP used in this study was Grade C-500 xGnPTM obtained from XG Sciences© (Lansing, MI, USA). GNP was heated at 1800°C for eight hours in a vacuum furnace (Solar Atmospheres, PA, USA) to remove functional groups, graphitise the surface of the GNP and produce vacuum annealed GNP (VA-GNP). VA-GNP-PTFE film was prepared by mixing defunctionalised VA-GNP with polytetrafluoroethylene, PTFE (60% w/w in ethanol) solution (Sigma-Aldrich) at a ratio of 9:1. Evaporation of ethanol left a homogeneous VAGNP-PTFE dense slurry which was rolled out into a cohesive, freestanding film. The resulting film was used for cytokine removal assessment to establish its efficacy as a haemoadsorbent.

A human hepatic epithelial cell line HepG2 (CRL-11997™, ATCC®, VA, USA) was used for the assessment of GNP cytotoxicity, with silver nanoparticles as a positive control.
The cytokine marker adsorption profile of GNP was first evaluated by incubating 10% w/v of GNP with fresh frozen human blood plasma (Cambridge Bioscience Ltd., Cambridge, UK) spiked with 1ng/mL of IL-8, IL-1β, IL-6, IL-10 and TNF-α (BD Biosciences, UK) for 60 min.

Results and Discussion

The porosity of GNP-based materials is summarised in Table 1.

Table 1. Porosity of GNP, VA-GNP and VA-GNP-PTFE film

Sample	S_{BET} (m^2/g)	V_{QSDFT} (cm^3/g)	D (nm)	d (nm)
Initial GNP	797	0.807	0.785	547
VA-GNP	294	0.397	0.723	1670
VA-GNP-PTFE	210	0.301	0.852	--

Specific surface area (S_{BET}), pore volume (V_{QSDFT}) and pore size (D) were calculated from low temperature nitrogen adsorption. Particle diameter(d) was determined using dynamic light scattering. The cytokine marker adsorption profile of GNP showed rapid and efficient removal of cytokines from human plasma spiked with a cytokine cocktail (Fig. 1A).

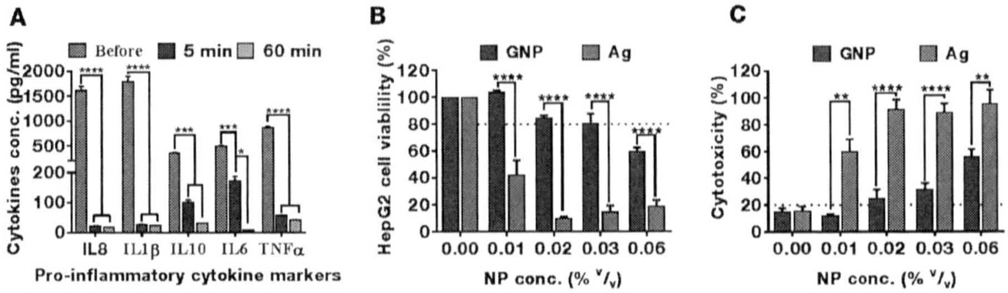

Figure 1. (A) the removal efficiency of inflammatory cytokine markers in spiked human plasma samples; GNP cytotoxicity assessments after 24 hours direct contact using (B) MTS and (C) LDH assays. ** represents $p<0.01$, *** represents $p<0.001$ estimated using two way ANOVA statistic test (±standard error of mean, n=3)

The LDH and MTT cytotoxicity assay results indicated that at lower tested concentrations, GNP caused no or minimal cytotoxic effect (< 20% cytotoxicity and >80% cell viability) towards HepG2 cells, while the silver nanoparticles were cytotoxic at concentrations as low as 0.01% (v/v) (Fig. 1B & C). A reduction in HepG2 cell viability after 0.06% GNP treatment to below 60%, indicating a suppression of metabolic activity (Fig. 1B). The cytotoxicity of GNP determined using LDH assay appeared to be dose-dependent and was significantly ($p<0.01$) lower than the silver nanoparticles at the same concentration (Fig. 1C). The slight cytotoxic effects of GNP at the concentrations of 0.03% and 0.06% are potentially due to the sheer volume of GNP in the two-dimensional culture system, blocking cell access to vital oxygen and nutrients.

Conclusions

We have demonstrated the use of GNP as a fast and very efficient adsorbent with low cytotoxicity for rapid removal of pro-inflammatory cytokines including IL-8, IL-1β, IL-6, IL-10, and TNF-α from human plasma on a clinically relevant scale. The material reported in this study shows faster adsorption compared to all other carbons reported so far, combined with a very high sorption capacity. Furthermore, we also developed a flexible freestanding GNP-PTFE composite material with high accessible surface area for targeted adsorption of cytokines and showed its potential as hemoperfusion adsorbent. For producing the composite material reported here, PTFE was used not only to produce a flexible freestanding film structure, but also to hold the GNP particles in place which could eliminate fines release often responsible for poor haemocompatibility of blood contacting materials. The non-cytotoxic nature of the GNP has been established in this study, however further research focusing on ensuring that this adsorbent does not trigger complement cascade, excessive platelet adhesion/activation or granulocyte activation is required.

Acknowledgment

This work was supported by the NOMAD project funded by the British Council and the UK Department for Business, Innovation and Skills through the Global Innovation Initiative.

Hydrolysis of cellulose catalyzed by oxidized carbon powder

Akira Takatsuki[1], Hirotomo Nishihara[1], Masahiko Hiratani[2] , Masato Kaneeda[2], Yoshiaki Kurihara[2], Takashi Kyotani[1]

[1] *Tohoku Univ.: Institute of Multidisciplinary Research for Advanced Materials, Sendai, Japan*

[2] *Hitachi Chemical Co.,Ltd.: Advanced Technology R&D Center, Tsukuba, Japan*

Presenting author's e-mail: takatsu@mail.tagen.tohoku.ac.jp

Introduction

The conversion of cellulose to valuable compounds is one of the important challenges in chemical industry. This reaction is typically performed with homogeneous catalysts such as sulfuric acid. As more eco-friendly and inexpensive alternatives, heterogeneous carbon-based catalysts functionalized with sulfonic groups[1)] have been developed. Moreover, Fukuoka et al. have demonstrated that oxygen-functional groups of carbon may also catalyze the reaction[2)]. Oxygen-functionalized carbons can be easily prepared with inexpensive methods and are expected as practical catalysts. However, the detailed mechanism as well as active sites has not been clarified yet. In this work, we have prepared carbon catalysts by dry-ball-milling and examined their effect on the cellulose conversion reaction. It has been revealed that the pulverization of the catalyst and some of thermally stable oxygen-functional groups are responsible for catalyzing the hydrolysis of cellulose.

Experimental

Gramax (Osaka Gas Chemicals), a soft carbon for anodes of Li-ion batteries, was used throughout this study. Fig. 1 shows a preparation scheme of carbon-powder catalysts. Gramax was crushed into fine powder by ball milling in air and automatically oxidized during the milling process. The resulting sample (M-Gramax) has some amount of oxygen-functional groups. M-Gramax was then heat-treated at 500 °C or 800 °C in Ar to desorb some functional groups and to prepare carbon catalysts with different surface chemistries as shown in Fig. 1. We denoted each heat-treated sample as M-Gramax-500, and M-Gramax-800, respectively.

Cellulose hydrolysis reaction was performed by heating the mixture of ball-milled cellulose (0.324 g), carbon catalyst (0.05 g), and distilled water (40 mL) at 180 °C for 1 h. The conversion ratio of cellulose was determined from the weight decrease of the solid phase by the reaction, while the yields of products were determined by the liquid chromatography of the aqueous phase.

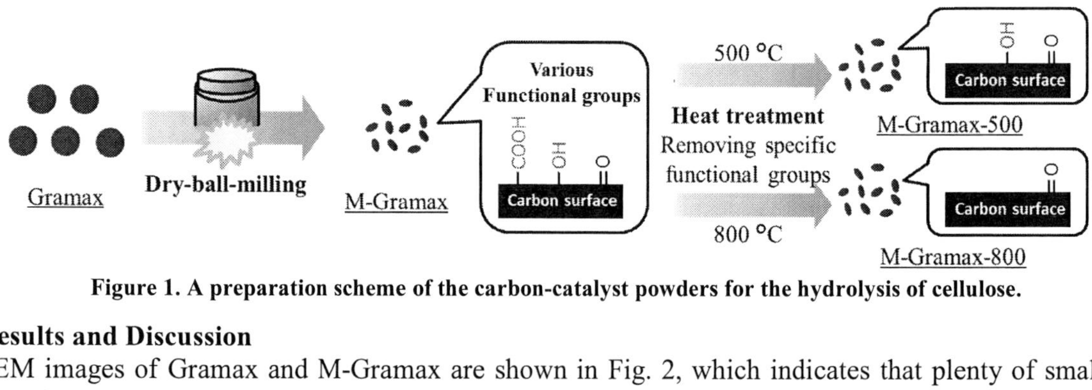

Figure 1. A preparation scheme of the carbon-catalyst powders for the hydrolysis of cellulose.

Results and Discussion

SEM images of Gramax and M-Gramax are shown in Fig. 2, which indicates that plenty of small particles with a diameter of less than 0.5 μm were produced from the pristine (unmilled) Gramax by the dry-ball-milling process. Furthermore, the structure of the catalysts was examined by a N_2 adsorption technique at -196 °C, revealing that the BET surface area of Gramax increased from 3 to 189 m^2/g by the ball-milling process. Both results evidence the pulverization of Gramax, which would be beneficial for good contacts between catalyst and substrates.

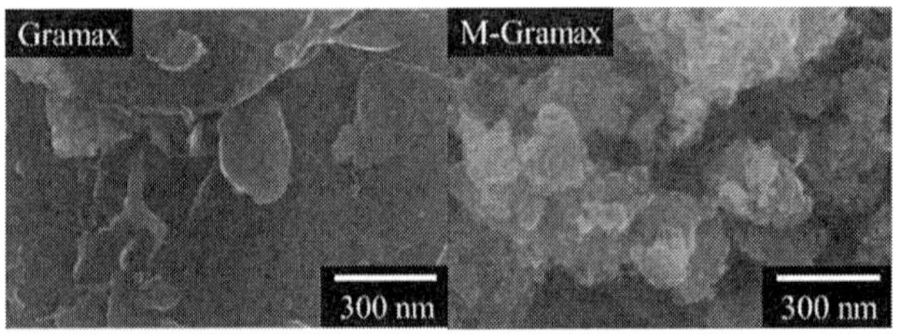

Figure 2. SEM images of Gramax and M-Gramax

Fig. 3 shows the yields of oligomers, glucose, and other by-products on the different carbon catalysts. A control experiment without catalyst gives only 32% conversion of cellulose with 14% yield of glucose, and the pristine Gramax exhibit only slight enhancement. The ball-milled product, M-Gramax, shows 57 % conversion of cellulose with 36 % yields of glucose. These results clearly indicate that the ball-milling process in air drastically enhances the catalytic activity of the parent carbon-powders. Interestingly, M-Gramax-500 retains the high catalysis to a great extent, while M-Gramax-800 completely loses its catalytic activity. The heat treatment at 500 or 800 °C does not change the surface area of the carbon powder, but does change both the number and the types of surface oxygen-functional groups as shown in Fig. 1. The catalysis enhancement by the ball-milling should be therefore derived not only from the sample pulverization but also from the activity of specific oxygen-functional groups which are thermally decomposed between 500 and 800 °C.

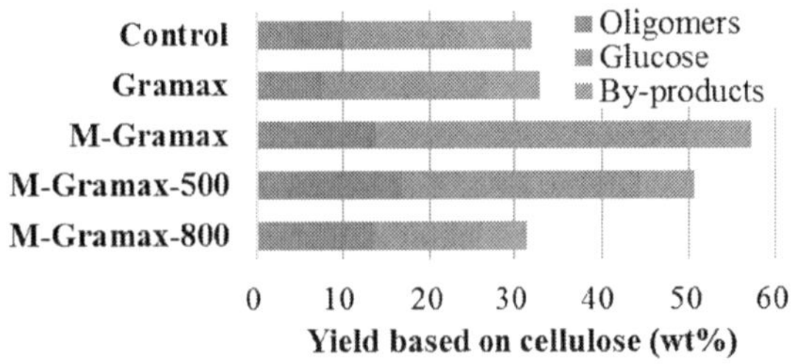

Figure 3. Yields of the products formed by the cellulose hydrolysis

To obtain deeper insight, the surface chemistry of carbon catalysts was analysed by temperature programmed desorption (TPD), in which the evolutions of CO and CO_2 were quantitatively measured as shown in Fig. 4. Unlike the pristine Gramax, M-Gramax shows noticeable gas evolutions. As expected, CO and CO_2 evolutions below 500 and 800 °C are remarkably decreased in M-Gramax-500 and M-Gramax-800, respectively. Since M-Gramax-500 retains a high catalysis, the oxygen-functional groups decomposed below 500 °C (e. g. carboxylic groups and acid anhydrides) are not responsible for the catalysis. Considering the little catalysis of M-Gramax-800, the CO-releasing groups which are stable at 800 °C (e.g. stable carbonyl groups) are not related to the catalysis. In conclusion, the oxygen-functional groups which are decomposed between 500-800 °C (e.g. phenol groups) should be the active sites of the cellulose dehydration catalysis.

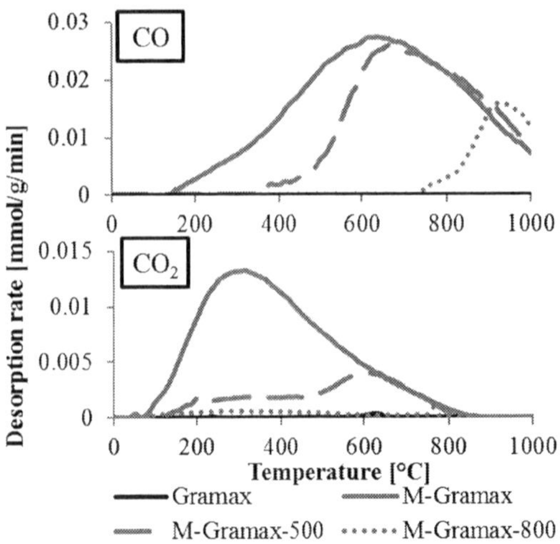

Figure 4. TPD profiles of the carbon catalysts.

Conclusions

Ball-milled carbon powders function as highly active catalyst for the hydrolysis of cellulose. The catalytic activity for the reaction can be attributed to both the pulverization of the carbon catalyst and some of the oxygen-functional groups which are formed on the carbon surface during the milling process.

References

1. Suganuma, Kiyotaka Nakajima, Masaaki Kitano, *et al.* (2008). Hydrolysis of Cellulose by Amorphous Carbon Bearing SO_3H, COOH, and OH Groups, *J. Am. Chem. Soc.*, 130, 12787-12793
2. Hirokazu Kobayashi, Mizuho Yabushita, Tasuku Komanoya, *et al.* (2013). High-Yielding One-Pot Synthesis of Glucose from Cellulose Using Simple Activated Carbons and Trace Hydrochloric Acid, *ACS Catal.*, 3, 581–587

SUSTAINABLE CARBON MATERIALS DERIVED FROM KRAFT BLACK LIQUOR

G. Amaral-Labat[1*], S.F. Quirino[2], M.R. Baldan[2], E.L. da Silva[3], A. Cuña[4], C.F. Malfatti[3], R.L. Marcos[5], G.F.B. Lenz e Silva[1]

[1]*Affiliation1: PMT. USP, São Paulo, Brazil*
[2]*Affiliation2: LAS. INPE, São Jose dos Campos, Brazil*
[3]*Affiliation3: LAPEC. UFRGS, Rio Grande do Sul, Brazil*
[4]*Affiliation4: DETEMA. Universidad de la República, Montevideo, Uruguay*
[5]*Affiliation5: Biophotonic. UNINOVE, São Paulo, Brazil*

*Presenting author's e-mail: gisele.amaral@usp.br

Introduction

The growing consummation of natural resources in the world has been stimulated industries and research centers to promote alternatives for avoiding this problem. Therefore, several researches are focused on the development of sustainable solutions, always aiming to reduce the environmental impact by the excessive use of resources. For this reason, the processes and products developed in the last decades have given greater attention for recycling by-products and wastes generated over the industrial processes. Indeed, the purposes in the utilization of by-products as raw materials for others industrial process consists in a decreasing of general expenses in new processes and final products. From this point of view, the black liquor, a residue rich in lignin and produced in large proportions in the pulp and paper industries, normally applied as an undervalued waste, starts to generate an interest in science and technology. The Kraft pulping process represents 90% of the world pulping production[1], generating a huge amount of liquid waste per year[2]. Hence, this work showed an alternative methodology to produce porous carbon materials from the raw black liquor.

Materials and Methods

Porous carbon materials were easily synthesized by chemical polymerization of the raw black liquor in alkaline medium using poly (methyl methacrylate) polymer (PMMA) in spherical format, as a "template" for the development of macroporosity and subsequent carbonization of the material in an inert atmosphere. The final porous materials presented low density (0.36 g/mL), high porosity (78%) and good thermal and mechanical stability. Porous carbon materials were also synthesized with nickel nanoparticles.

Results and Discussion

Porous carbon material with tunable properties, such as morphology (figure 1), porosity and composition were obtained by the different ways of preparation.

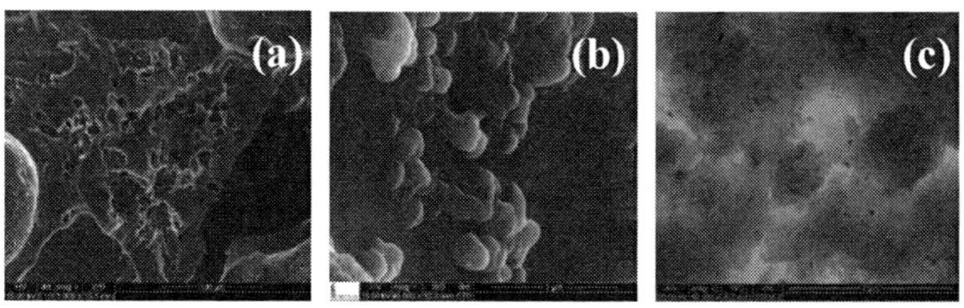

Figure 1. SEM images of three different carbon materials synthesized from the raw Kraft black liquor: (a) developed micro porosity; (b) nano-structured; (c) porous carbon as support of nickel nanoparticles.

The difference between the carbon porous samples are relevant for application in several fields. In the present work synthesized materials were tested (i) as biomaterial in bone regeneration, (ii) in electromagnetic shielding systems and (iii) as catalyst support for direct ethanol fuel cells (DEFC).

The viability and cytotoxicity of the porous material were evaluated *in vitro* with an osteoblasts culture at 24, 48 and 72 hours, and the results showed both parameters similar to the control (composed of the osteoblasts cells without material). Furthermore, cells were clearly seen surrounding the carbon material (figure 2(a)), and no halo was observed. The results strongly suggests that porous material may be used as a substrate for osteoblasts cells as confirmed by the number of cells calculated from viability and cytotoxicity methods.

Carbon material having controlled pore and particles sizes was mixed with phenolic commercial resin for preparation of composites (C@P). Samples with 2mm of thickness and two different percentage of carbon on resin matrix were tested as electromagnetic shielding, for aerospace application, using a vector network analyser (VNA) in the frequency range from 8.2 GHz to 12.4 GHz (X-Band). The results showed a high potential for application as shielding absorbed material (SE_A) composite. Figure 2b shows SE_A as a function of frequency. From the experimental measurement, the SE_A shows a dependence with the carbon material content presented values between 7-9 dB for samples with a larger content of carbon material.

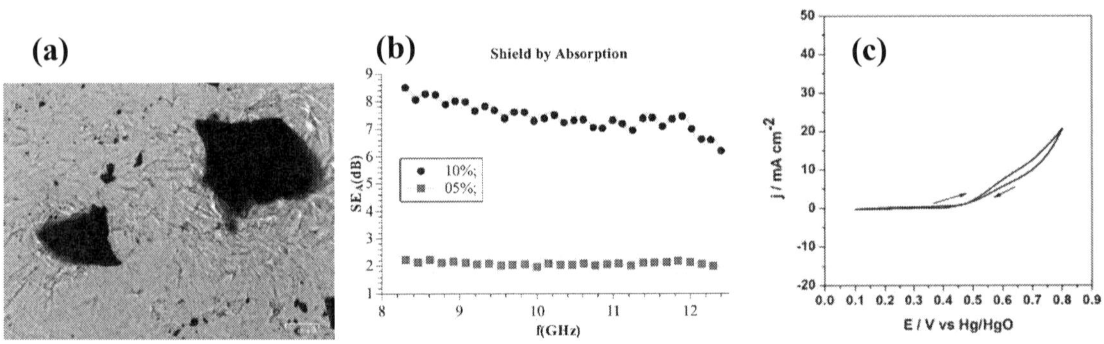

Figure 2. (a) Optical microscopy of carbon material 72 hours after osteoblasts cells growth (40 μm); (b) Electromagnetic shielding by adsorption for C@P (5 and 10 % of carbon) in the frequency of the X-Band ; (c) Cyclic voltammogram of Ni/C in a solution of 1.0 mol L^{-1} NaOH + 1.0 mol L^{-1} ethanol.

Finally, the material showed as an efficient catalyst support for ethanol electro-oxidation for fuel cells. The carbon material supported Nickel nanoparticles (Ni/C) and the catalytic activity was evaluated by cyclic voltammogram of the sample at 25°C in 1.0 mol L^{-1} NaOH, presenting specific current density higher than 20 $mA.cm^{-2}$ with a scan rate of 10 $mV.s^{-1}$.

Conclusions

The proposed methodology shows innovation based on the production of porous carbon materials regarding sustainability concerns and it might present multiple advantages for the pulp and paper industry as (i) increasing the added value to the raw waste, (ii) increasing in the production of pulp and paper, and (iii) the feasibility of using the raw Kraft black liquor without the generation of new by-products because the steps of purification of lignin are not required. Furthermore, the employed methodologies of synthesis are suggested as promising techniques in the production of different porous carbon materials since the final product might be applied for several applications.

References

1. Azadi, P., Inderwildi, O.R., Farnood, R., King, D.A. (2013). Liquid fuels, hydrogen and chemicals from lignin: A critical review. *Renewable & Sustainable Energy Reviews* 21, 506-523.

2. Gouveia, S.L., Fernández-Costas, C., Sanromán, M.A., Moldes, D. (2013). Polymerization of Kraft lignin from black liquors by laccase from Myceliophthora thermophila: Effect of operational conditions and black liquor origin. *Bioresource Technology* 131, 288-294.

MECHANISM OF CNTS GROWTH: UNDERSTANDING 3 KINETIC ROUTES, CHIRALITY AND SHAPE

Luis Sousa Lobo,
Requimte Research Center, Univ. Nova, Campus da Caparica, Caparica, 2825 Portugal
sousalobo266@gmail.com

Introduction

"The central problem in in nanotube science is still the mechanism. Although progress has been made, the growth mechanism of both single- and multi-walled nanotubes remain controversial", remarked Peter Harris[1]. The difficulties on the essence of the mechanism remain, despite much recent progress on controlling and optimizing shape and rate[2-6]. Following a presentation at the Carbon 2016 conference and a perspective article in the journal Carbon[7], we developed the details of the chirality effects in 2 of the 3 routes of the proposed mechanism, considering the shape and solid-state nature of the nanoparticle "sides" operating and so explaining the growth and structure of the nanotubes. We followed a remark by Donald Bethune about explaining chirality using our approach. Our proposal for the CNTs and CNFs mechanism is in line with the observed kinetics – detailed kinetics - which seems to be the best working base to understand the mechanism[8,9].

Solid-state chemistry, geometry at nanoscale and C atoms flux operating

Bulk diffusion of C atoms in transition metals is relatively easy due to the ratio of the atomic radius of solute and solvent being less than 0.60: dissolution of C is interstitial. C surface diffusion is not competitive. Once at reaction temperature, initial solid-state changes of the catalyst may occur, as well as after the admission of the reactant gas - vg initial Ni_3C formation using Ni [2]. When graphene nucleation is successful a steady-state C flux and growth rate is be established. Fick's 1st law operates. When the C bulk diffusion is the rate-controlling step the order of reaction is zero and the activation energy (dependence on T) is the same as E_a for C diffusion in the metal used as catalyst[8,9]. Nucleation and growth take place in different regions of the catalyst particle, with different crystal orientation (check Fig. 1)[9]. Different sizes and shapes of the nanoparticles facilitate different forms of growth. The time found for 2 mm "penetration" of C in Fe metal blocks was 61 hours[10]. So we can estimate that the time required in a nanoparticle with a diameter of 20 nm is 2 seconds.

Specific growth modes: a) Zig-zag vs armchair was found to change with the shape of the catalyst nanoparticle[11]. This is explained in Fig. 2; b) Spiral growth occurs when at the base of the two sides of a CNT growing the C flux is asymmetrical: a pentagon is formed due to the inclination of the growing tube in the faster growth side and an heptagon on the slower growing side; c) Helical growth may occur when an asymmetrical nanoparticle is operating with the successive carbons growing forming hexagons with the half hexagon turning always to the same side due to a non-symmetrical geometry, instead of alternating as it happens in symmetrical growing.

References

1. Peter Harris, Carbon Nanotube Science, Cambridge UP, 2011.
2. De Jong K, Geus JW, Catal. Rev. Sci. Eng. 42 (4) (2000) 481-510.
3. Lamouroux E, Serp, P, Catal. Rev. Sci. Eng. 49 (3) (2007) 341-405.
4. Zang J, Terrones M et al. Carbon 98 (2016) 718-32.
5. Tessonier V, Bichara C. Carbon 58 (2013) 2-39-
6. Nessim GD et al. Carbon 49 (2011) 804-10.
7. Lobo LS,.Carbon 114 (2017) 411-417
8. Lobo LS, Trimm DL. Nature Physical Science 1971
9. Lobo LS. Reac Kinet Mech Cat (2016) 118: 393-414
10. Mehl RF. J. App. Physics, 8, 174 (1937) 174-85.
11. Koziol, KKK et al. Chem. of Materials (2010) 22, 4904-11

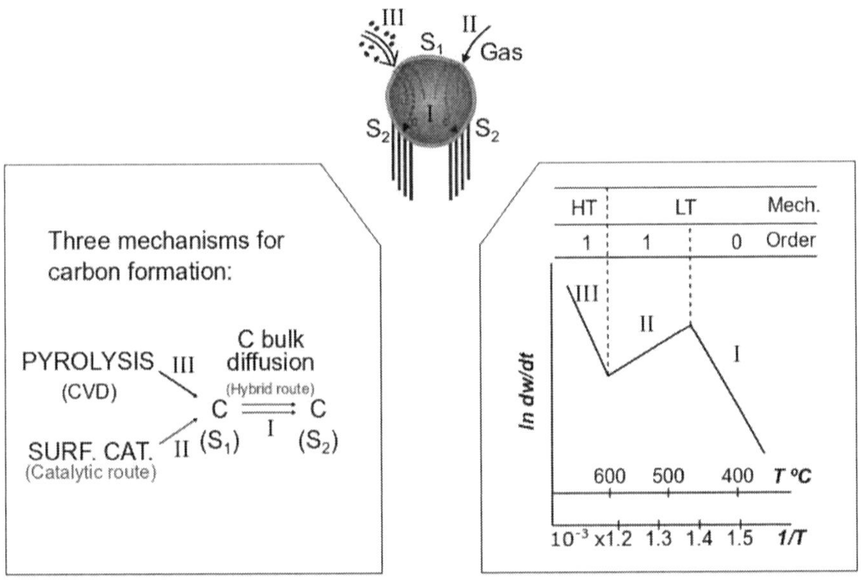

Fig. 1. Interpreting the Arrhenius plot. Alternative routes in C formation

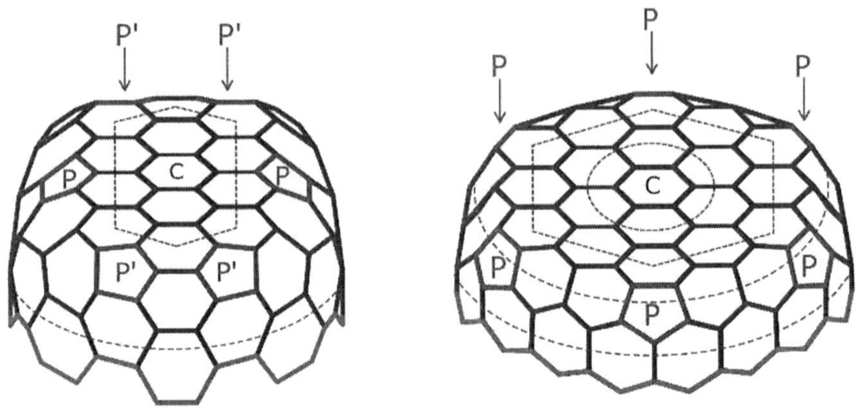

Fig. 2. Zig-zag (spherical np) vs armchair (oval np) CNT growth

ANALYSIS OF BOUND POLYMER IN CARBON-RUBBER INTERFACE BY USING CARBON-COATED ALUMINA NANOPERTICLES AS A MODEL FILLER

Yasuto Hoshikawa[1], Rei Kawaguchi[1], Hidenobu Akahane[2], Takafumi Ishii[1], Norihisa Hoshino[1], Tomoyuki Akutagawa[1], Hiroshi Yamada[2] , Takashi Kyotani[1]

[1] *Tohoku University, Institute of Multidisciplinary Research for Advanced Materials (IMRAM), Sendai, Japan*
[2] *Tire Materials Advanced Development Department Department, Kodaira, Japan*

*yasuto.hoshikawa.e4@tohoku.ac.jp

Introduction

When carbon black (CB) is mixed with rubber, a part of the rubber polymers is bound on the CB surface by physisorption and/or chemisorption (such polymers are called "bound rubber"), and as a result, the mechanical strengths of the CB/rubber composites have been significantly improved. Although it is well known that such reinforcement effect strongly depends on the state of CB/rubber interface, a detailed structure of the rubber-filler interface has not been revealed yet because it is very hard to extract only the information of the external carbon surface directly interacting with the rubber molecules. We have found that loosely bound rubber surrounding the CB filler can be completely removed when the composite is autoclaved in toluene at 150 ºC[1], but there is still a large amount of inner carbon in the CB filler, which hampers the detailed analysis of the rubber-filler interface.

In this context, we have realized that alumina nanoparticles (ANPs) are useful as a model filler to understand the interface because the secondary structure of ANPs is similar to that of CB and, more importantly, the external surface can be covered with a very thin carbon layer; i.e., the filler does not contain the inner carbon, thereby making it possible to analyze only the external carbon. Herein, we completely cover ANPs with a few graphene layers by chemical vapor deposition (CVD) and then prepare the rubber composite with the resulting carbon-coated ANPs (C-ANPs) as a filler, followed by the removal of loosely bound rubber by the autoclave (AC) treatment. Finally, we attempt to analyze the bound polymers formed in the rubber-filler interface with various means.

Materials and Methods

The ANPs (14 nm in an average primary size) were carbon-coated by acetylene (at 600 ºC) or methane (at 900 ºC) CVDs, followed by a heat treatment at 800, 1000 or 1150 ºC in N_2. Hereafter the samples were named XY or XY_Z (X: CVD gas (A = acetylene, M = methane), Y: CVD temperature, Z: heat treatment temperature). The C-ANPs were characterized by thermogravimetry (TG) in air, TEM observation, N_2 adsorption at -196 °C and the temperature programmed desorption (TPD) method. The C-ANPs are mixed with styrene-butadiene rubber (SBR) by a banbury mixer, and the resulting composite was autoclaved in toluene at 150 ºC to remove the soluble rubber components including loosely bound rubber (the samples are named XY_Z_AC). The amount of bound rubber remained in the C-ANPs was calculated by subtracting the weight decrease upon the TG measurement up to 800 ºC in air from the weight of carbon deposited on the C-ANPs. In order to remove the alumina, the AC-treated C-ANP was immersed into hydrofluoric acid (HF) (the samples are named XY_Z_AC_HF), and then a glass transition temperature (T_g) of the remaining SBR was measured by differential scanning calorimetry (DSC).

Results and Discussion

Among the pristine ANPs and the C-ANPs, the values of BET specific surface area (S_{BET}) are similar (120~140 m^2/g). This is also the case for total pore volume (V_t). The findings indicate that the structure of the ANPs doesn't basically change even upon the carbon deposition by the CVD. The amounts of carbon on the C-ANPs (W_c) are also almost the same (ca. 14 wt%) for all samples, and the number of theoretical graphene layer calculated from the W_c and the S_{BET} is approximately two.

TEM images of A600 and M900_1150 are shown in Figure 1, where the alumina nanoparticles covered with carbon layers are observed. The number of the carbon layer estimated from the images is almost the same (2 - 4 layers), but the two samples significantly differ in the structure of the stacking carbon layers on the ANPs. In the case of A600, the length of the carbon layers is short and the stacking layers are somewhat disordered. On the other hand, the carbon layers of M900_1150 look continuous and orderly stacked, clearly indicating that the degree of crystallinity of M900_1150 is higher than that of A600.

Hydrogen (H) contents in the carbon deposited on the C-ANPs were estimated by the TPD method and it is found that the H content is decreased with increasing the CVD and/or the heat treatment temperatures. A600 and M900_1150 have the highest and the lowest H contents (12.1 and 1.2 mmol/g), respectively, and the H contents of the other three samples are in between. Thus, we have succeeded in preparing the five types of the C-ANPs with almost the same carbon amount but with different amounts of hydrogen contents. Since all hydrogen atoms should be bonded with carbon edge sites, the present C-ANPs differ in the number of hydrogen-terminated edge sites.

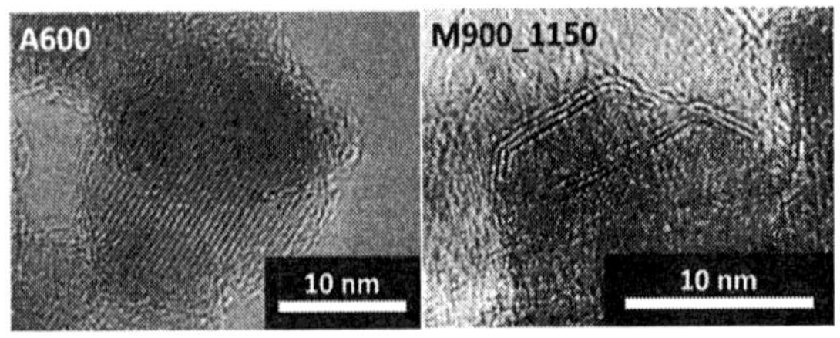

Figure 1 TEM images of C-ANPs samples (A600 and M900_1150).

Next, we discuss a relationship between the bound rubber formation in the carbon-rubber interface and the amount of hydrogen in carbon edge sites. As mentioned in Introduction, the AC treatment can remove most rubber components except for the rubber polymers strongly bound in the interface [1]. Figure 2 shows plots of the amount of such bound polymers in the interface against the number of H atom in the external carbon layer for all samples. When the H content is less than 1.15×10^{18} pcs/m^2, the amount of bound polymer is linearly increased. In this range, any H terminated edge sites on the external carbon surface can react with polymer radicals generated by mixing with filler. Accordingly, the bound polymers become the chemisorbed ones. However, when the number of H is more than 1.15×10^{18} pcs/m^2, the amount of bound polymer levels off and becomes a constant (1.25 mg/m^2). One of the possible explanations is as follows. All polymer radicals are already consumed by the H atoms of 1.15×10^{18} pcs/m^2 and therefore chemisorbed polymers aren't formed any more. Because the formation of the physisorbed polymer must be independent of the H terminated edge sites, an intercept of the linear line of the chemisorbed polymer corresponds to the amount of physisorbed polymer [1].

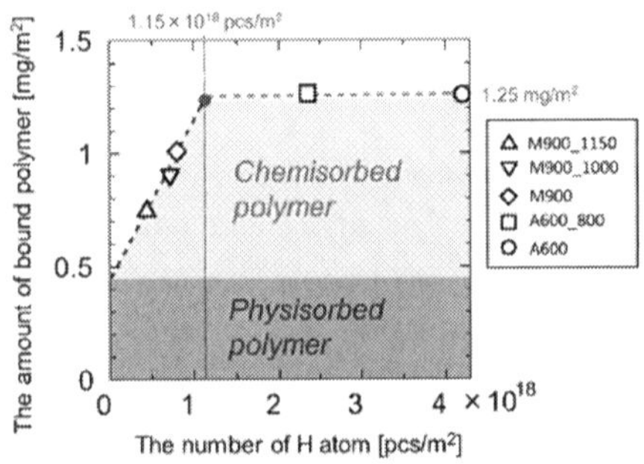

Figure 2 Plots of the amount of bound polymer against the number of hydrogen (H) atoms in the external carbon layer of the C-ANPs

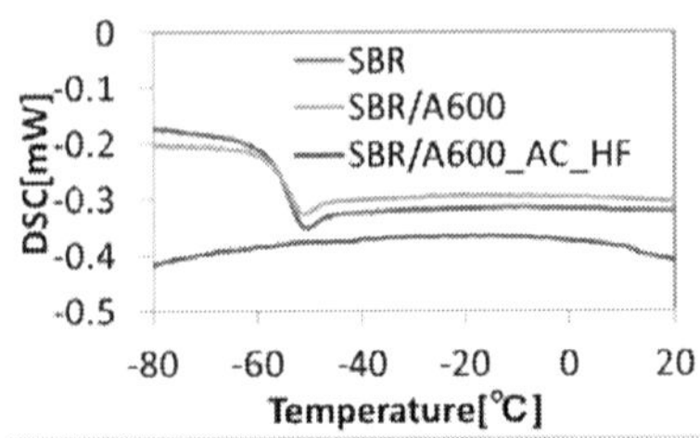

Figure 3 DSC curves of pristine SBR, SBR/A600 composite and SBR/A600_AC_HF samples.

Figure 3 shows DSC curves of the pristine SBR, the SBR/A600 composite and the SBR/A600_AC_HF samples. The former two samples exhibit a sudden decrease to the endothermic direction from -55 °C, which is T_g of the SBR. In contrast, there is no endothermic decrease in the curve of the SBR/A600_AC_HF sample; *i.e.*, the absence of T_g in the bound polymers on the SBR-A600 interface, This finding suggests that the molecular motion of side chains in the SBR polymers is strongly restricted even at more than the T_g of the pristine SBR. In other words, the glassy state is kept at room temperature. It is noteworthy that ,in the cases of conventional SBR/CB composites and the SBR/A600 sample, it is impossible to detect such a specific feature of the bound polymers because a fraction of the bound polymer is too small, compared with all other components (carbon, unbound rubber, loosely bound rubber and alumina). Thus, by using the unique model filler (C-ANP), we have confirmed the glassy state of bound rubber polymer with DSC for the first time.

References

1. Hoshikawa, Y., An, B., Kashihara, S., Ishii, T., Ando, M., Fujisawa, S., Hayakawa, K., Hamatani, S., Yamada, H., Kyotani, T.. (2016). Analysis of the interaction between rubber polymer and carbon black surfaces by efficient removal of physisorbed polymer from carbon-rubber composites. *Carbon* 99, 148-156

INFLUENCE OF THE DEGREE OF OXIDATION OF GRANULAR ACTIVATED CARBON ANODES IN ARSENATE ELECTROADSORPTION

Luis Felipe Cházaro-Ruiz[1*], Rigoberto Santoyo-Cisneros[1], José René Rangel-Méndez[1], Mario Alberto González-Lara[1]

[1]*División de Ciencias Ambientales, Instituto Potosino de Investigación Científica y Tecnológica, A.C., Camino a la Presa San José 2055, Col. Lomas 4a. Sección, C.P. 78216 San Luis Potosí, SLP, México*

*luis.chazaro@ipicyt.edu.mx

Introduction

The presence of arsenic in groundwater has become a serious worldwide problem due to the high toxicological issues. Electroadsorption processes using activated carbon as electrode materials have become an interesting alternative for the removal of arsenate ions from groundwater.[1-3] However, many questions about which activated carbon properties have impact during arsenate electroadsorption have arisen. The objective of this work was to study the influence of the chemical surface of granular activated carbon of bituminous origin on arsenate electroadsorption.

Materials and Methods

The activated carbon material used was granular F400 of bituminous origin. This material was either chemically oxidized with nitric acid (F400 Ox) or thermally reduced (F400 T) under H_2/Ar atmosphere. These materials were characterized by Boehm titrations and physisorption of N_2, and subsequently used to prepare electrodes, using stainless steel as a current collector. The electrodes were immersed in arsenate solutions (2.5 mg L^{-1}) and used as anodes, in batch experiments under constant stirring.

The specific capacitance of the materials was estimated in 1 M H_2SO_4, by cyclic voltammetry at 1 mVs^{-1}. The experimental arrangement for electroadsorption studies consisted of a system with 3 electrodes, in a 250 mL glass cell. The reference electrode was an Ag/AgCl/KCl(sat.) system. The auxiliary electrode was a stainless steel mesh and the working electrodes (WE) were stainless steel mesh with activated carbon. The initial concentration of As(V) was determined from an adsorption isotherm. The experiments were carried out in two stages: I) The electrodes were exposed to the arsenic solution without the application of an electric potential, until reaching the adsorption equilibrium. II) Then, anodic potentials (0.4, 0.8 and 1.2 V *vs* Ag/AgCl/KCl(sat.)) were applied after reaching this adsorption arsenate equilibrium with no potential. These potentials depended on the working potential window of the cell given by the type of anode used. Likewise, the different anodic potentials were applied from the beginning (t=0 h) in order to study this effect on the kinetics of electroadsorption.

Results and Discussion

The capacitance values obtained for F400 and F400 T were similar to those reported in the literature for granular activated carbon materials in acid media, which are in a range from 80 to 120 Fg^{-1}. The voltammetric responses were typical rectangular forms of electrical double layer capacitor. The voltammetric response of the oxidized F400 Ox presented wide peaks attributed to faradaic reactions which do not favor the electroadsorption process.

When the potentials were applied in stages, the electroadsorption capacity of the thermally treated carbon was higher than the pristine F400 due to its greater electrical conductivity and higher specific capacitance. On the other hand, its electroadsorption speed increased with the application of potential from the beginning, and also required lower potential than the pristine material, according to what is observed in **Figure 1** and **Table 1**.

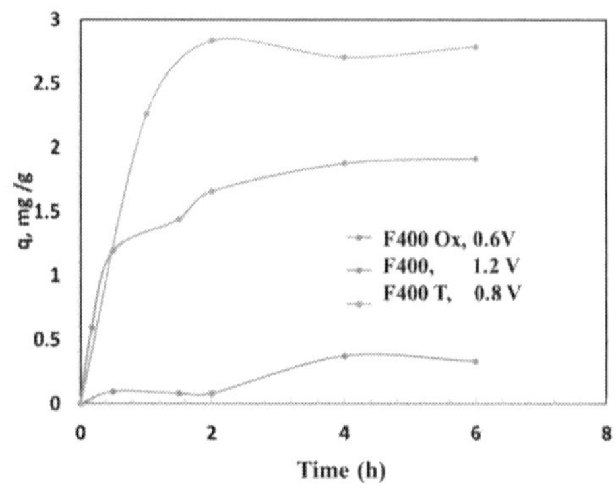

Figure 1. Electrosorption of As(V) using anodes of pristine F400, thermally trated (F400 T), and oxidized (F400 Ox). Experimental conditions: $C_{As(V)}$ 2.5 mg L^{-1}, T 25 °C, pH=7.0, at 300 rpm.

Table 1. Electroadsorption capacities of As(V) of the anodes

Anode	Anodic potential (*vs* Ag/AgCl(KCl, sat)	q_t (mg. g^{-1})
F400 OX	0.6 V	0.33
F400	1.2 V	1.91
F400 TT	0.8 V	2.79

During the presentation the results of electrochemical impedance spectroscopy will also be discussed. In this case the material with greater capacity of electroadsorption of arsenate was the one (F400 T) with more basic character (E_{pzc}=11.8), which despite of having lower hydrophilicity had a good performance as electric double layer capacitor. In contrast, the oxidized material, whose adsorption capacity did not increase when using polarization. The F400 T presented the highest arsenate removal by electroadsorption (70%) in comparison with F400 (55%) after adsorption with no potential and the gradual increase of electric potential.

Conclusions

The study of the surface chemistry properties of activated carbon materials used in electroadsorption processes is fundamental to understand the mechanisms involved in the adsorption of arsenate by polarizing activated carbon materials. The effects of anodic polarization on activated carbon materials are not only reflected in an increase in the adsorption capacity, but in the speed with which the process is carried out.

Acknowledgment

The authors express their gratitude to CONACYT for the financial support granted through Atención a Problemas Nacionales project No. 789. The technical support of the Q.F.B. Elizabeth Cortez, M. C. Dulce Partida and M. C. Guillermo Vidriales; CONACYT for doctorate scholarship No. 423679. Also, we thank M. del Carmen Rocha from National Laboratory LAMBAMA.

References

1. Bain, E. J., Calo, J. M., Spitz-Steinberg, R., Kirchner, J., Axon, J. (2010). Electrosorption/electrodesorption of arsenic on a granular activated carbon in the presence of other heavy metals. *Energy and Fuels* 24, 3415–3421.

2. Beralus, J.-M., Ruiz-Rosas, R., Cazorla-Amorós, D., Morallón, E. (2014). Electroadsorption of Arsenic from Natural Water in Granular Activated Carbon. *Frontiers in Materials* 1, 1–9.

3. Fan, C.-S., Tseng, S.-C., Li, K.-C., Hou, C.-H. (2016). Electro-removal of arsenic(III) and arsenic(V) from aqueous solutions by capacitive deionization. *Journal of Hazardous Materials* 312, 208–215.

IONOTHERMAL NANOPOROUS CARBONS: SUPERIOR ADSORBENTS FOR THE REMOVAL OF DICLOFENAC AND SULFAMETHOXAZOLE FROM SYNTHETIC WASTEWATER

Ana S. Mestre[1,2*], Raquel Gonçalves[1], Valentine Vigneron[1], Stephanie Lemaire[1], Cristina Freire[2], Ana P. Carvalho[1]

[1] *Centro de Química e Bioquímica and Centro de Química Estrutural, Faculdade de Ciências, Universidade de Lisboa, Portugal*
[2] *REQUIMTE/LAQV, Departamento de Química e Bioquímica, Faculdade de Ciências, Universidade do Porto*

*asmestre@fc.ul.pt

Introduction

Water treatment technologies improvement is the key to face current water quality challenges related with the presence of pollutants in (waste)waters, e.g. pharmaceutical compounds (PhCs) that resist to conventional treatments. Although powdered activated carbons (PACs) are considered one of the best available technologies to control PhCs, the use of commercial PACs in conventional (waste)water treatment plants presents drawbacks related with their limited kinetics and adsorption under real environmental conditions.

The present work aims to evaluate the influence of micro-mesopore network of a set of PACs prepared by ionothermal route on the removal of two PhCs – diclofenac and sulfamethoxazole – in synthetic wastewater. Since most adsorption studies are focused on evaluating the adsorption of PhCs onto PAC using deionized water[1] these data will allow a step forward to understand the PhCs adsorption mechanisms in a system more close to the real scenario, i.e. synthetic organic-free wastewater.

Materials and Methods

Ionothermal nanoporous carbons were prepared from glucose using eutectic mixtures of $LiCl/ZnCl_2$, $NaCl/ZnCl_2$ and $KCl/ZnCl_2$, designated, respectively, by LiZn, NaZn and KZn, and also only with $ZnCl_2$ for comparison purposes. Briefly, 1 g of glucose was mixed with 3 g of the salt mixture and activated under N_2 flow (5 cm^5 s^{-1}) with the following heating scheme: from room temperature up to 240 °C (10 °C min^{-1}) for 2 h, further increased up to 800 or 1000 °C (10 °C min^{-1}), and kept for more 2 h.[2] Sample was washed with distilled water until no detection of chlorine ions in the supernatant and dried. Characterization of the carbon materials was made by SEM, N_2 and CO_2 adsorption at, respectively, -196 and 0 °C, and by the determination of the pH at the point of zero charge (pH_{PZC}).

PhC adsorption onto synthesized PACs and a material commercialized by Cabot-Norit for water treatment purposes (Norit SAE Super, NS) was evaluated by kinetic and equilibrium assays. The inorganic aqueous matrix was prepared with NaCl, $CaCl_2$ and $NaHCO_3$ and the pH was adjusted to 7.1 by adding HCl/NaOH to resemble the inorganic background of a wastewater treatment plant (WWTP) secondary effluent.[3] For kinetic assays the initial concentration of each PhCs was 2.5 mg/L and 10 mg/L of PAC was used. Two target PhCs were spiked in the test waters, diclofenac/DCF (anionic, relatively hydrophobic) and sulfamethoxazole/SMX (anionic, hydrophilic), selected for their distinct hydrophilic/hydrophobic character. PhCs were quantified by UV-vis spectroscopy.

Results and Discussion

Ionothermal carbons have a total pore volume higher than 1 cm^3 g^{-1} and A_{BET} values from 1054 to 2302 m^2 g^{-1} presenting well developed micro and mesopore networks (Table 1). There is always a decrease of BET area and micropore volumes and an increase of mesopore volume as the radius of the alkaline metal in the eutectic mixture increases. Comparing the textural properties of material G/Zn/1000 prepared by $ZnCl_2$ activation with those of ionothermal samples it is possible to attribute the increase of mesopore network to the presence of the alkaline metals during the activation process. Pore size distributions (Figure 1) reveal that materials prepared with Zn and LiZn are those presenting the highest amount of small mesopores (2 – 5 nm) while KZn favoured the development of mesopores with larger widths (10 – 40 nm). Regardless the carbon source, and eutectic mixture, the preparation yields varied between 20 and 35 %.

Table 1. Textural properties of the ionothermal and commercial PACs

Sample	A_{BET} m²/g	V_{total}[a] cm³/g	V_{meso}[b] cm³/g	$V_{\alpha\ micro}$ cm³/g	$V_{\alpha\ ultra}$ cm³/g	$V_{\alpha\ super}$ cm³/g
G/Zn/1000	2302	1.44	0.28	1.16	0.00	1.16
G/LiZn/1000	2156	1.54	0.62	0.92	0.00	0.92
G/NaZn/1000	1389	1.01	0.65	0.36	0.00	0.36
G/KZn/1000	1054	1.27	1.02	0.25	0.09	0.16
NS	1070	0.79	0.36	0.43	0.00	0.43

[a] Evaluated at p/p^0 = 0.975 in the N_2 adsorption isotherms at -196 °C
[b] Difference between V_{total} and $V_{\alpha\ total}$

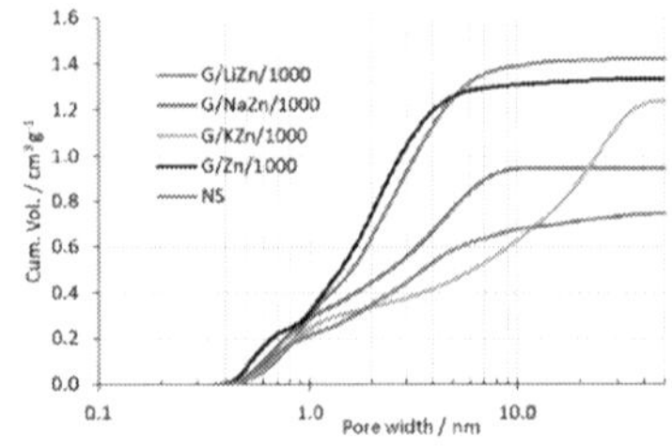

Figure 1. Cumulative pore size distributions obtained from the 2D-NLDFT-HS method applied to the N_2 adsorption isotherms

Kinetic assays (Figure 2) reveal that materials presenting more developed micro-mesopore networks attain higher SMX and DCF uptakes. In all the cases SMX is less prone to adsorb than DCF. Kinetic data fitted to the pseudo-second order kinetic model and equilibrium isotherm data has better fitting to the Langmuir equation. Equilibrium data shows that material G/LiZn/1000 attains twice the maximum adsorption capacity of NS carbon for both targets, and has also higher affinity towards the adsorptives.

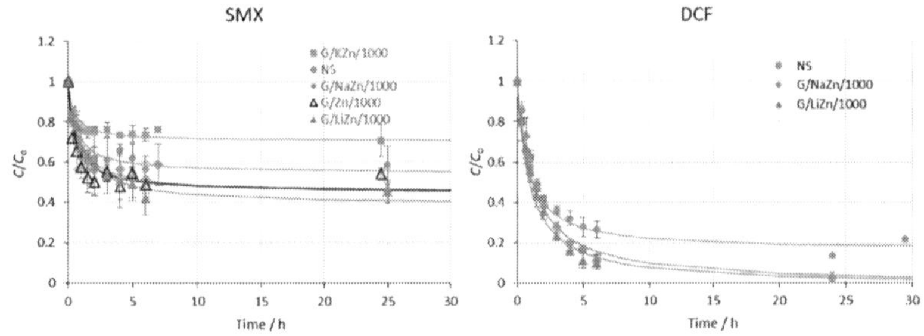

Figure 2. PhCs adsorption kinetics from inorganic synthetic water (10 mg/L PAC and 2.5 mg/L each PhC)

Conclusions

Ionothermal carbons presenting high amount of small mesopores outperform the commercial golden standard for the removal of the target pharmaceuticals in synthetic inorganic matrix spiked at the low ppm level. The lower removal efficiencies and affinities of sulfamethoxazole for the tested carbons is certainly related to its hydrophilic character.

Acknowledgment

The authors acknowledge financial support of projects UID/MULTI/00612/2013 (CQB) and UID/ QUI/50006/2013 - POCI/01/0145/FERDER/007265 (REQUIMTE) from FCT/MEC through national funds and co-financed by FEDER, under the Partnership Agreement PT2020. ASM thanks FCT for a Post-doctoral grant SFRH/BPD/86693/2012. The authors thank Salmon & Cia for the supply of carbon NS.

References

[1] Mestre, A.S., Tyszko, E., Andrade, M.A., Galhetas, M., Freire, C., Carvalho, A.P. (2015). Sustainable activated carbons prepared from a sucrose-derived hydrochar: remarkable adsorbents for pharmaceutical compounds. *RSC Advances*, 5(25), 19696-19707.
[2] Zhang, Y., Yu, B., Liu, Z. (2014). Highly mesoporous carbons derived from biomass feedstocks templated with eutectic salt $ZnCl_2$/KCl. *Journal of Materials Chemistry* A 2(45), 19324-19329.
[3] Rodriguez, E., Campinas, M., Acero, J.L., Rosa, M.J. (2016). Investigating PPCP Removal from Wastewater by Powdered Activated Carbon/Ultrafiltration. *Water, Air, & Soil Pollution* 227(6), 177.

CARBON ELECTRODE MODIFICATION FOR EXTENSION OF THE CAPACITOR VOLTAGE

Elzbieta Frackowiak*, Adam Slesinski, Krzysztof Fic

Poznan University of Technology: Institute of Chemistry and Technical Electrochemistry, Poznan, Poland

*Presenting author's e-mail: elzbieta.frackowiak@put.poznan.pl

Introduction

Activated carbons are the most common materials used for electrochemical capacitors (ECs) operating in aqueous medium as well as in organic electrolytes (**Fig. 1**). The effect of carbon porosity, matching of size pores with ions, conductivity of carbons and electrolytes, the role of different cations and anions, their soft and hard character, and also their solvation force should be considered.
The scientific interest for ECs functioning in water based solutions is growing intensively. It is well known that energy of device is proportional to the square of operating voltage. Therefore, the investigations on voltage window extension over the thermodynamic stability of water (1.23 V) is of high importance.

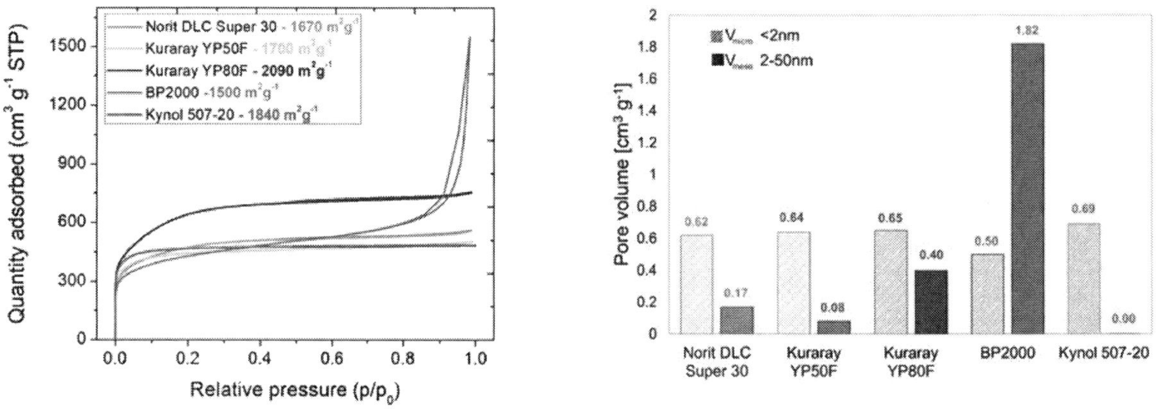

Fig. 1 Common activated carbons used for electrochemical capacitor

Increase of the capacitor voltage can be realized by different ways, e.g.

- application of neutral electrolytes (sulfates, nitrates, iodides, etc.),
- selecting of two different electrolytes for positive and negative electrodes,
- self-buffering of both electrode/electrolyte interfaces by ammonia treatment.

Neutral salts as capacitor electrolytes

The most often electrolytic aq. solutions used for electrochemical capacitor are Li_2SO_4, K_2SO_4, KNO_3, $Mg(NO_3)_2$, KI, NaI, KSCN etc. with their various concentrations. The type of carbon materials used as EC component in various electrolytes has a significant importance for the voltage range. As a rule of thumb, carbon materials rich in defects, edge-sites provoke electrolyte decomposition.
It is noteworthy that some of electrolyte ions can supply additional pseudocapacitive reactions, eg. iodides, pseudohalides. Beneficial role of such electrolytes on electrochemical capacitor performance was proved.

Different electrolytes for (+) and (-) electrodes

A lot of efforts was devoted to hybrid system where both electrodes operate in electrolytes with different pH. Separation of both electrolytic solutions by a buffer plays a crucial role. **Fig. 2** shows effect of combination two different electrolytes and comparison with organic medium. Energy values for aqueous and organic electrolytes are comparable.

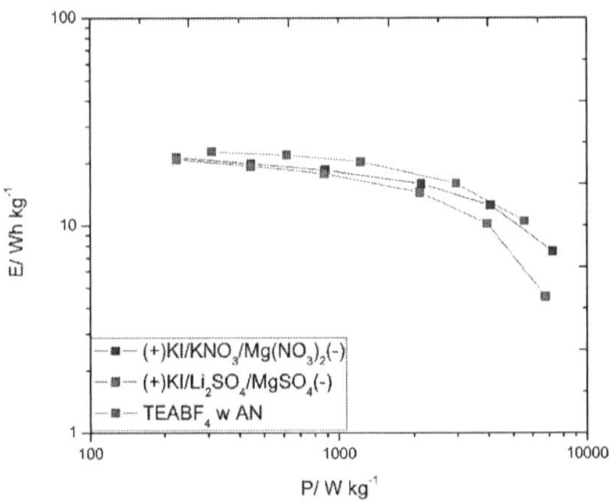

Fig. 2 Ragone plot for hybrid system [1]

Self-buffering by ammonia treatment

It has been proved that ammonia species can strongly modify the pH of positive and negative electrodes, in turn, the EC voltage range is safely extended [2]. Carbon electrodes have been treated by ammonia at room temperature. Even if a moderate amount of nitrogen was introduced into carbon, the physicochemical analysis evidenced various functional groups on the carbon surface. It appears that oxidized carbon has a better affinity to ammonia, in turn, the various character of surface functionality is observed (Table 1).

Table 1. Surface functionalities character of carbon electrodes before and after NH_3 treatment

	before ammonia treatment		**after ammonia treatment**	
	acidic groups mmol g^{-1}	basic groups mmol g^{-1}	acidic groups mmol g^{-1}	basic groups mmol g^{-1}
k50	0.16	0.17	0.71	0.2
k50ox	0.42	0.25	0.58	0.38
k50ox300	0.33	0.38	0.45	0.43
k50ox500	0.3	0.43	0.41	0.46

Shift of the rest potential of the electrodes towards more negative was a clear proof that the modification was done effectively. Positive electrodes have been protected against overoxidation. Such ammonia treated electrodes reveal an excellent long-term cyclability of electrochemical capacitor.

Ageing phenomena of carbon electrodes

Electrochemical capacitor should operate thousands of cycles. Such long-term operation causes significant modification/degradation of carbon electrode morphology, structure and elemental composition. Ageing phenomenon of carbon electrode can be monitored by *in-situ* or *ex-situ* experiments. *In-situ* measurements, which utilize special techniques such as gas analysis (CO, CO_2, NO, N_2, H_2 etc.) by mass spectrometry allow to monitor carbon electrode during its operation. Generally, only one electrode can be investigated. Thermoprogrammed desorption coupled with mass spectrometry supplies additional information about surface functionality of carbon electrode after cycling (*ex-situ* measurements). Especially, positive electrodes have been enriched in oxygen functional groups. Very precious information can be obtained by *post mortem* analysis of electrodes (nitrogen gas sorption at 77K, Raman spectroscopy, elemental analysis etc.). It reflects real changes of carbon after long-term cycling such as modification of specific surface area, micro/mesopore volume, sp^2/sp^3 ratio, chemical composition.... Both electrodes are investigated separately and compared with pristine material what gives full information on ageing phenomena.

Conclusions

- Extension of operating voltage until 1.8V was successfully realized by different ways.
- Ageing mechanism for both electrodes is different, strongly dependent on the carbon material, electrolyte type.
- Positive electrode can be oxidized but without oxygen evolution.
- Ammonia treatment of carbon electrodes causes self-buffering of pH at both electrodes.

Acknowledgment

The authors would like to acknowledge the financial support from the grant of Ministry of Science and Higher Education, no. 03/31/DSPB/0353.

References

1. Menzel, J.; Fic, K.; Frackowiak, E. (2015). Hybrid aqueous capacitor with improved energy/power performance *Progress in Natural Science: Materials International* 25, 642-649
2. Slesinski, A.; Matei-Ghimbeu, C.; Fic, K.; Béguin, F.; Frackowiak, E. (2018). Self-buffered pH at carbon surfaces in aqueous supercapacitors. *Carbon* 129, 758-765

Formatting Issue - Best Available Paper/Graphic

CONSISTENCY OF CARBON PSD CHARACTERIZATION METHODS BASED ON 2D-NLDFT MODELS APPLIED TO ADSORPTION DATA OF AR, N_2, AND O_2

Jacek Jagiello* and Jeffrey Kenvin

Micromeritics Instrument Corporation, Norcross, Georgia, USA

*jacek.jagiello@micromeritics.com

Introduction

The pore size distribution (PSD) of porous carbons is most often derived from the analysis of nitrogen and argon adsorption isotherms measured at their boiling temperatures. From the two gases argon is regarded as more reliable[1] for the PSD analysis because of its minimal specific interactions with the surface chemical groups. Such interactions may influence adsorption of nitrogen molecules that possess a significant quadrupole moment and thus may interact specifically with the surface polar groups.

In practice, however, using liquid argon as a cryogen for argon adsorption measurements may be challenging due to its high cost and a limited availability in various parts of the world.

For the purpose of the carbon PSD characterization we propose using oxygen as an alternative to nitrogen. The quadrupole moment of oxygen is about 3.5 times lower than that of nitrogen, which makes oxygen significantly less susceptible to specific interaction than nitrogen. It is fundamentally and practically important that oxygen is a liquid at 77 K.

Materials and Methods

In this work we use Supelco microporous carbon samples and the original PX-21 Amoco (KOH activated) carbon sample. Adsorption isotherms of nitrogen and oxygen were measured at 77 K and argon at 87 using the high-resolution Micromeritics 3Flex instrument equipped with high-vacuum system, three micropore ports and three 0.1 Torr pressure transducers. The samples were degassed under vacuum at 620 K overnight prior to the adsorption measurements.

Results and Discussion

For the analysis of the adsorption isotherms we use the 2D-NLDFT[2-3] models (kernels) developed for these gases. The theoretical adsorption kernel isotherms were calculated using Tarazona's version[4-5] of the non-local density functional theory (NLDFT). Following our earlier works[2-3] we assume that carbon pore walls can be described as corrugated graphene sheets where the profile corrugation is given by a simple sine curve. In addition, we introduce a function F that describes the variation of the adsorption energy sites[3].

The corrugation parameters are obtained by fitting the theoretical 2D-NLDFT isotherm to the experimental O_2 adsorption data measured on the nongraphitized carbon black BP 280. Using these parameters we calculated the 2D-NLDFT kernel that constitutes our oxygen adsorption model for carbons with corrugated pores.

To test the new model we analyzed the adsorption isotherms of O_2 and N_2 at 77 K and Ar at 87 K measured on the reference microporous carbon C1012 from Supelco (Fig. 1) and on PX-21 from Amoco (Fig. 2). All models provide excellent fits to the experimental data. The fitting plots are free from artifacts, such as S-shape deviations, that are known for the results obtained with the standard NLDFT models. In both cases presented in Fig. 1b and Fig. 2b the PSDs calculated from all three gases are in a very good agreement. It is important to note that particularly the PSD results for argon and oxygen are in the quantitative agreement.

Formatting Issue - Best Available Paper/Graphic

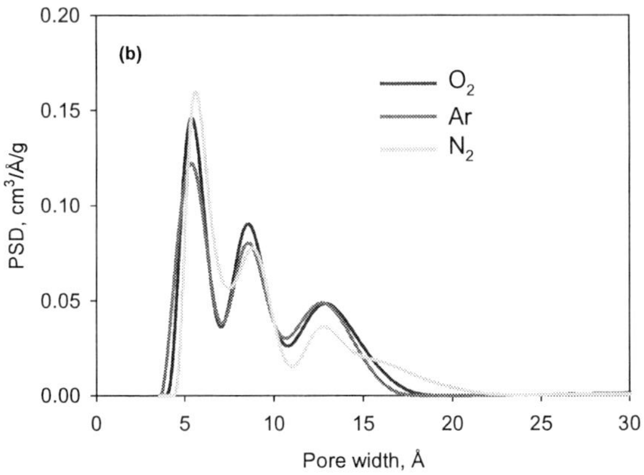

Figure 1. PSD Analysis of C1012 carbon (Supelco) from isotherms of O_2 and N_2 at 77 K and Ar at 87 K using 2D-NLDFT models. (a) Fit of the models to the experimental isotherms. (b) Comparison of the calculated PSDs from three gases.

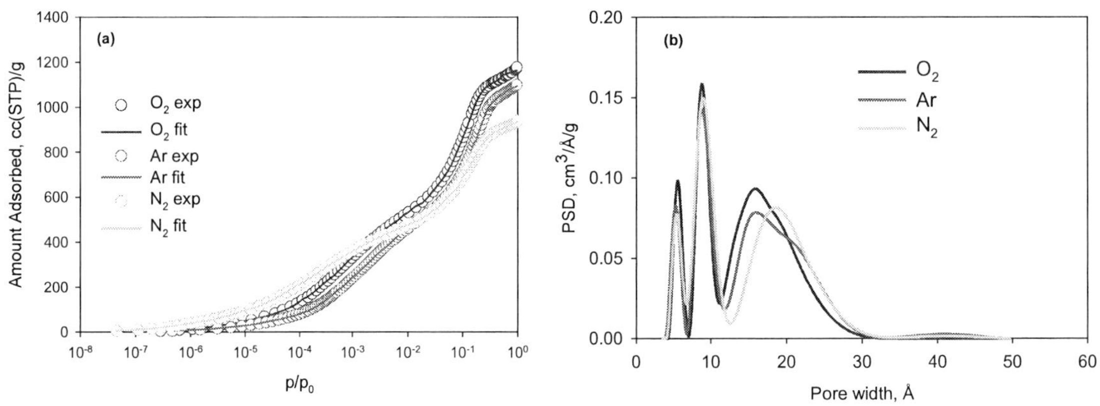

Figure 2. PSD Analysis of PX-21 carbon from isotherms of O_2 and N_2 at 77 K and Ar at 87 K using 2D-NLDFT models. (a) Fit of the models to the experimental isotherms. (b) Comparison of the calculated PSDs from three gases.

Conclusions

The presented results indicate that the oxygen adsorption at 77 K may be a good replacement for argon at 87 K. An additional benefit of using O_2 at 77 K is its low saturation pressure of 155 mmHg which reduces the measurement error related to the free space correction.

Acknowledgements

We would like to thank Dr. Ani Nikova of Cabot Corporation for providing a sample of carbon BP280, Dr. William Betz of Sigma-Aldrich Corporation for providing Supelco samples of carbon molecular sieves and Dr. Matthew Kulzick of BP America for providing PX-21 Amoco carbon.

References

1. Thommes M. *et al.* Pure and Applied Chemistry **87**, 1051-1069 (2015).
2. Jagiello J., Olivier J. P. Carbon **55**, 70-80 (2013).
3. Jagiello J., Olivier J. P. Adsorption **19**, 777-783 (2013).
4. Tarazona P. Physical Review A **31**, 2672-2679 (1985).
5. Evans R., Marconi Marini Bettolo U., Tarazona P. J Chem Phys **84**, 2376-2399 (1986).

ADVANCES ON NOBLE METAL-FREE ELECTROCATALYSTS BASED ON CARBON FOR ENERGY CONVERSION DEVICES

D. Sebastián[1*], G. Lemes[1], J.M. Luque-Centeno[2], M.V. Martínez-Huerta[2], E. Pastor[3], M.J. Lázaro[1]

[1] *Instituto de Carboquímica (CSIC), Zaragoza, Spain*
[2] *Instituto de Catálisis y Petroleoquímica (CSIC), Madrid, Spain.*
[3] *Instituto Universitario de Nanomateriales y Nanotecnología, Departamento de Química, Facultad de Ciencias, Universidad de La Laguna, La Laguna, Spain.*

*Presenting author's e-mail: dsebastian@icb.csic.es

Introduction

Carbon nanostructures have received increasing attention in electrocatalysis due to their surface electronic configuration with abundant π electrons, which are beneficial for electron demanding reactions like those occurring in electrochemical energy conversion devices. This is the case of fuel cells, which efficiently and directly convert chemical energy into electrical energy, representing the ideal solution for non-polluting and highly efficient vehicles and power consumers. The sluggishness of the oxygen reduction reaction (ORR), taking place at the cathode of fuel cells, limits the entire process. It requires the utilization of platinum-based electrodes, which scarce resources and high cost represent the main concern for the implementation of this kind of system at large scale.

Several strategies have been pursued to replace platinum.[1] Among them, doped carbon materials are a more recent and unexplored alternative with great perspectives in terms of cost and availability, but yet poor activity and stability. In particular, graphene structures are expected to allow for the enhancement of catalytic activity derived from its peculiar 2D morphology. There is still a great and recent controversy about the nature of active sites and the influence of non-noble metals like Fe or Co on the electro-catalytic activity. In any case, the presence of low atomic weight atoms, like nitrogen, as dopants in the catalyst matrix is essential to confer appropriate oxygen adsorption characteristics and sufficient catalytic activity.

In this work, we will review our most recent strategies to develop highly efficient graphene-based and noble metal-free electrocatalysts and the derived results in terms of fuel cell performance.

Materials and Methods

Graphene oxide (GO) was synthesized according to modified Hummer's method by exfoliation of graphite powder (<20 μm, Sigma-Aldrich). Nitrogen doped samples were prepared via thermal annealing of a mixture of GO and urea under inert atmosphere at different temperatures (500 °C, 600 °C, 700 °C and 800 °C) for 2 hours. Metal composites of Co or Ti with N-doped reduced graphene oxide (rGO) were synthesized by a sol-gel methodology, followed by thermal annealing under inert atmosphere at 800 ºC.[2]
The as-synthetized catalysts were physico-chemically characterized by X-ray diffraction (XRD), X-ray photoelectron spectroscopy (XPS), elemental analysis (EA), and inductively coupled plasma - mass spectrometry (ICP-MS).
The electrocatalytic activity towards ORR was determined in a three-electrode system, using a glassy carbon rotating disk electrode (RDE) in alkaline medium (0.1M NaOH).

Results and Discussion

All GO samples were reduced to rGO upon thermal treatment, with an average number of layers below 10, as confirmed by XRD experiments (not shown). XPS analyses (not included) indicated the presence of pyridinic, pyrrolic and quaternary N, as commonly reported in the literature.[3]

The introduction of nitrogen in the graphene matrix creates a positive effect on ORR activity, as depicted in Figure 1 (in terms of half-wave potential). There is still an open discussion about understanding the correlation between the nature of active sites and ORR performance of this kind of porous carbon-based catalysts.[4] An optimum nitrogen doping procedure at high annealing temperatures results in active materials for ORR in alkaline environment. On the other hand, it appears that a large content of nitrogen does not necessarily improve activity and can even have a negative influence. Carbon nitride phases were individuated within the reduced graphene structure from the thermal decomposition of urea below 600 ºC. Carbon nitride, although exhibiting a larger density of C-N species than N-doped rGO, shows poorer electrical conductivity.[5] Other than N, the presence of transition metals (Co and Ti in our case) may also contribute to the positive shift of potential for ORR. Concerning composites with Ti and Co metals, the best results were achieved thanks to synergy effects of N-doped rGO and Co/Co_2O_3, benefiting the active site formation and facilitating the high performance for oxygen reduction reaction.

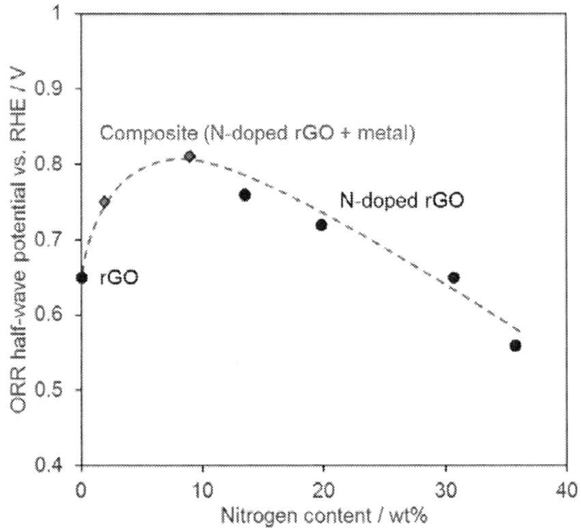

Figure 1. ORR half-wave potential as a function of nitrogen content in rGO-based catalysts (metal = Co or Ti).

Conclusions

Noble metal-free catalysts can be designed from carbon materials, such as graphene, by doping strategies including low atomic weight elements, nitrogen being the most appealing, or transition metals with appropriate surface structure and chemistry. The synthetic pathway greatly influences the physico-chemical features that ultimately promote the attainment of active electrocatalysts.

Acknowledgment

Authors acknowledge financial support given by Spanish Ministry of Economy and Competitiveness (MINECO) through projects ENE2014-52158-C2-1-R and 2-R (co-founded by FEDER). J. M. Luque and G. Lemes also thank MINECO and Aragon Government, respectively, for their Ph.D. grants. D. Sebastián thanks also MINECO for his Ramón y Cajal research contract (RyC-2016-20944).

References

1. Shao, M., Chang, Q., Dodelet, J.P., Chenitz, R. (2016) Recent advances in electrocatalysts for oxygen reduction reaction. *Chemical Reviews* 116, 3594-3657.

2. Luque-Centeno, J.M., Martínez-Huerta, M.V., Sebastián, D., Lemes, G., Pastor, E., Lázaro, M.J. (2018). Bifunctional N-doped graphene Ti and Co nanocomposites for the oxygen reduction and evolution reactions. *Renewable Energy* 125, 182-192.

3. Wood, K., O'Hayre, R., Pylypenko, S. (2014) Recent progress on nitrogen/carbon structures designed for use in energy and sustainability applications. *Energy & Environmental Science* 7, 1212-1249.

4. Zhu, C., Li, H., Fu, S., Du, D., Lin, Y. (2016) Highly efficient nonprecious metal catalysts towards oxygen reduction based on three-dimensional porous carbon nanostructures. *Chemical Society Reviews* 45, 517-531.

5. Zhang, Y., Ge, J., Wang, L., Wang, D., Ding, F., Tao, X., Chen, W. (2013) Manageable N-doped graphene for high performance oxygen reduction reaction. *Scientific Reports* 3, 2771.

Graphitization of activated carbon obtained from local biomass (babassu) and their electrochemical properties

Anupama Ghosh[1], Claudia do Amaral Razzino[2], Archi Dasgupta[3], Kazunori Fujisawa[4], Jose R. A. da Cunha[5], Anderson O. Lobo[2,6], Joshua Robinson[3], Mauricio Terrones[3,4], Bartolomeu C. Viana[1,6,*]
[1]*Departament of Physics, Federal University of Piaui, Teresina, PI, Brazil*
[2] *IP&D, Vale do Paraíba University, São José dos Campos, SP, Brazil*
[3] *Materials Science and Engineering, Pennsylvania State University, University Park, PA, USA*
[4]*Department of Physics, Pennsylvania State University, University Park, PA, USA*
[5]*Faculdade de Ciências Exatas e Naturais, Federal Univesrsity of Para, Cametá, PA, Brazil*
[6] *Program of Materials Science and Engineering, Federal University of Piaui, Teresina, PI, Brazil.*
*Presenting author's e-mail: *bartolomeu@ufpi.edu.br*

Introduction

New-generation supercapacitors and ion batteries are one of the most important research area dedicated to find out the novel smart and functional materials for this kind of applications[1-3]. Graphite has long been introduced to commercialize battery devices[1], activated carbon have also been investigated as electrode materials[4-6]. Although activated carbon (AC) meets the basic criteria of being a good supercapacitor electrode material, the major disadvantage is its non-crystallinity which limits the ion-diffusion. Indeed, graphitization is the most widely used method to improve the electrochemical performance of ACs along with chemical activation, halogenation and catalytic carbonization[7-10]. Although it dramatically improves the conductivity of ACs, the high degree of graphitization decreases the surface area and total pore volume drastically[11,12]. So an optimization is required between these factors to achieve the best possible results leading to a partial graphitization with somewhat retention of the surface area and pore structure[13]. Babassu coconut is a fruit of an abundant palm tree in north of brazil. Here, we have synthesized AC with elevated surface area and porosity from the part of this local biomass, mesocarp, using hydrothermal carbonization followed by chemical activation (KOH) at higher temperature and it was applied as electrode material. The sample structure obtained at different temperatures were characterized. Finally, all produced and heat-treated ACs were submitted to various electrochemical tests which revealed that all of them are useful for eletrochemical applications depending on the temperature of treatment.

Materials and Methods

It was used a tubular furnace and a flow of nitrogen (1000 sccm) to perfom pyrolysis and activation of biomass (KOH). The hydrothermal carbonization (pre treatment) was carried out in a Teflon-lined steel autoclave in a muffle furnace. Raman and IR spectroscopy were performed using a Bruker Senterra and Vertex 70 spectrometer, respectively. SEM was performed quanta FEG 250-FEI and the cyclic voltammetry (CV) measurements of ACs electrodes were performed on a model PGSTAT 128N Autolab (Eco Chemie). The electrochemical impedance spectroscopy (EIS) measurements were performed on a model PGSTAT 302N Autolab.

Results and Discussion

ACs obtained were submitted to heat treatment at 1100°C, 1400°C and 2200°C (AC-1100, AC-1400 and AC - 2200, respectively). AC-2200 undergone graphitization and it was confirmed by characterizations in Fig.1. In the Raman spectra the D', D", D and G bands, two amorphous bands were observed. It was possible to observe that the intensities of amorphous bands disappeared in AC - 2200. AC-2200 has graphitic behavior which has been confirmed by TEM, XRD and Raman characterizations. The CV of ACs was carried out in were performed in Faradaic mode. The data obtained from CV and EIS experiments were electroactive area of electrodes, double layer capacitance and charge transfer resistance are shown in Table 1. The AC-1400/GCE presents highest double layer capacitance and a great surface area. In contrast, the AC-2200/GCE exhibit a capacitance value of 95-fold smaller than AC-1400. Both are excellent electrodes for applications that require high surface area. The AC-1400/GCE presents high capacitance and can be used as an electrode of the electrical double layer capacitor. The AC-2200/GCE presents a low value for Rct and can be used as electrochemical sensors, because a large electroactive area together with the low Rct favors the detection sensitivity.

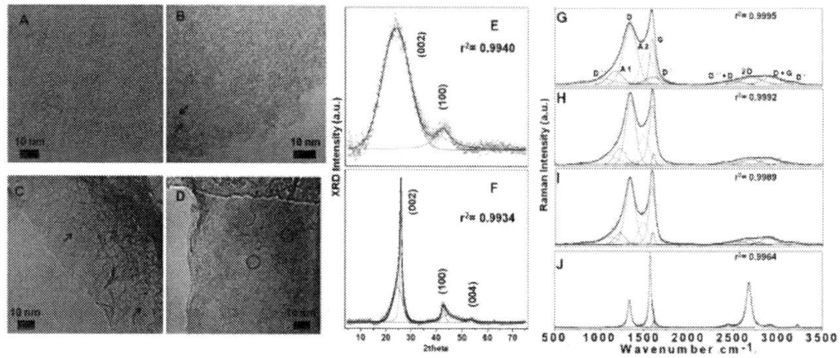

Figure 1. TEM micrographs of (A) AC, (B) AC-1100, (C) and (D) of AC-2200; XRD diffractogram of (E) AC and (F) AC-2200; Raman bands of AC, AC-1100, AC-1400 and AC-2200 using 532 nm (2.33 eV) laser source.

Table 1. Electroactive area, double layer capacitance and charge transfer resistance values of the electrodes. The electrodes have a geometric area of 0.0314 cm^2

Electrodes	*ΔEp (mV)*	*A_{EA} (cm^2)*	*Capacitance (mF cm^{-2})*	*R_S (Ω)*	*R_F (Ω)*	*R_{CT} (Ω)*	*OCP-DC (mV)*
GCE	73	0.0266	0.012	156	-	22.8	276
AC/GCE	221	0.23	18.3	157.2	516.4	50.19	222
AC-1100/GCE	160	0.27	33.0	157.4	157.6	59.76	224
AC-1400/GCE	196	0.42	105.3	158.7	2311	2975	226
AC-2200/GCE	54	0.23	1.1	158.7	-	22.47	266

Conclusions

It was produced ACs from a Brazilian biomass (babassu mesocarp) with different structures, morphologies and electrochemical applications. The AC-1400/GCE electrode presents high capacitance and can be used as an electrode of electrical double layer capacitor and the AC-2200/GCE presents a low value for Rct which can be used as electrochemical sensors.

Acknowledgment

Authors thank to the CAPES-PROAP and FAPEPI agencies for the funds supporting this oral presentation.

References

1. T. Kim, C. Jo, W. G. Lim, J. Lee, J. Lee, K. H. Lee, *Carbon* 104 (2016) 106-111.
2. Y. Tiana, Y. Songa, Z. Tanga, Q. Guo, L. Liu, *Journal of Power Sources* 184 (2008) 675-681.
3. X. Zhou, F. Chen, T. Bai, B. Long, Q. Liao, Y. Ren, J. Yang, *Green Chem.* 18 (2016) 2078-2088.
4. C. Portet, G. Yushin, Y. Gogotsi, *Carbon* 45 (2007) 2511–2518.
5. B. Chang, Y. Guo, Yanchun Li, H. Yin, S. Zhang, B. Yang and X. Dong, *J. Mater. Chem.* A 3 (2015), 9565-9577.
6. T. A. Silva, H. Zanin, E. Saito, R. A. Medeiros, F. C. Vicentini, E. J. Corat, O. Fatibello-Filho, *Electrochimica Acta* 119 (2014) 114-119.
7. A. R. Mohamed, M. Mohammadi, G.N. Darzi, *Renew. Sustain. Energy. Rev* 14 (2010) 1591.
8. P. Barpanda, G. Fanchini, G. G. Amatucci, *Carbon* 49 (2011) 2538-2548.
9. P. Barpanda, G. Fanchini, G. G. Amatucci, *J Electrochemical Soc*, 154 (2007) 467-476.
10. K. Inomata, Y. Otake, *J. Mater. Sci* 44 (2009) 4200.
11. Y. Hanzawa, H. Hatori, N. Yoshizawa, Y. Yamada, *Carbon* 40 (2002) 575.
12. I. Martin-Gullon, J. P. Marco-Lozar, D. Cazorla-Amoros, A. Linares-Solano, *Carbon* 42 (2004) 1339–1343.
13. Z. Xu, D. Cai, Z. Hu, F. Wang, L. Gan, *Electrochimica Acta* 117 (2014) 486– 491.

ADVANCED CARBON MATERIALS BY ELECTROSPINNING OF LIGNIN

José Rodríguez-Mirasol, Francisco J. García-Mateos, Tomás Cordero

Universidad de Málaga, Andalucía Tech, Departamento de Ingeniería Química, Campus de Teatinos s/n, 29010 Málaga, España

*mirasol@uma.es

Introduction

One-dimensional nano- and submicron-structured materials, such as fibers, tubes or wires, are of great technological and industrial interest, particularly for chemical engineering in the fields of energy storage, adsorption and catalysis, due to their unique physical and chemical properties. These fibrous materials show substantial improvements on heat and mass transfer, pressure drop and productivity under reaction conditions and on electrical properties in electrochemical devices as compared to powder or granular materials. Electrospinning is a simple and versatile technique for obtaining fibers and tubes of different nature, organic or inorganic, in the submicron- and nonoscale. This technique makes use of the electrohydrodinamic forces to produce nano- or submicrometric liquid jets from a nozzle much larger in diameter than the jets itself that solidify while it is still in its cylindrical, metastable form, producing a solid fiber. This communication deals with the study of preparation, characterization and functional applications of lignin-derived carbon fibers. The results obtained show that electrospinning of lignin, the second most abundant polymer in nature, is an advantageous and promising approach for the preparation of carbon fibers with interesting properties for applications in adsorption, catalytical and electrochemical processes.

Materials and Methods

Different carbon materials were prepared by electropinning of Alcell lignin/ethanol solutions in a coaxial configuration[1]. The effect of adding H_3PO_4 (weight ratio H_3PO_4/lignin = 0 and 0.3, respectively) to the initial electrospinning solution on the physical-chemical properties of final carbon materials was studied. Surface functionalization and/or depositing metal catalysts well dispersed on the lignin fibers may be done in a single step by easily adding the functional and/or the metal precursor to the electrospun solution. Lignin fibers were thermo-stabilized in air atmosphere at different heating rates (between 0.05 to 3 ºC/min) from 60 to 200 ºC. Subsequently, stabilized lignin fibers were carbonized/activated at 900 ºC.

Results and Discussion

The incorporation of H_3PO_4 to the initial lignin solution produces more oxidized spun lignin fibers, due to the reaction of phosphoric acid with the dissolved lignin, generating phosphate (and/or polyphosphate) esters throughout the structure of the lignin fibers. These phosphate groups seem to be the responsible for the production of cross-linking reactions during the stabilization step and enable to shorten the conventional stabilization process to avoid fiber fusion from more than 90 h to only 2 hours. Furthermore, incorporation of H_3PO_4 promotes the chemical activation of lignin fibers and greatly enhances their oxidation resistance, avoiding their complete combustion during carbonization under low concentration of O_2 at temperatures as high as 900 °C.[2] **Figure 1** shows the micrographs of different carbon materials prepared by electrospinning of lignin solutions. Varying the heating rate during stabilization step, carbon fibers with different degrees of interconnection were prepared (**Figure 1a and b**). Carbon tubes were prepared by electrospinning of lignin solution (**Figure 1c**) in a tri-axial configuration using glycerine a liquid template by the inner needle and well surface dispersed metal (Pt) carbon fibers were also prepared by adding the metal precursor to the initial solution (**Figure 1d**).

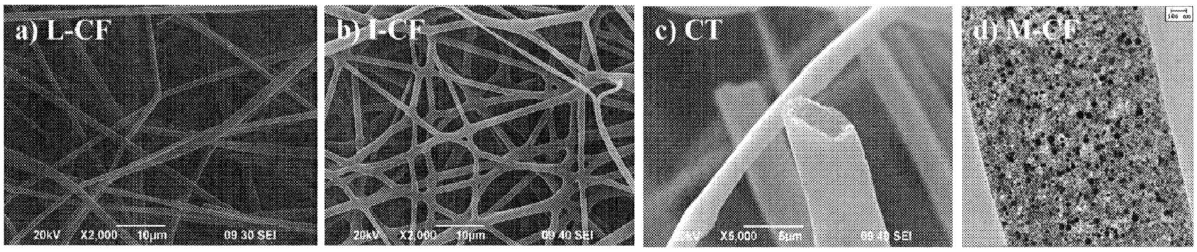

Figure 1. Carbon fibers/tube micrographs. a) Linear carbon fibers; b) Interconnected carbon fibers; c) Carbon tubes; d) metal-loaded carbon fibers.

Lignin-based activated carbon fibers resulted efficient as adsorbents for CO_2 capture under post-combustion conditions, obtaining a CO_2 adsorption capacity of 1.3 mmol/g at 15 kPa and 25 °C.[3] Phosphorus surface functionalization of the lignin-based carbon fibers allowed modifying the surface chemistry of the final fibrous material that showed acid surface properties and remarkable catalytic behavior in the dehydration of bioethanol and methanol to ethylene and dimethyl ether, respectively. P-containing carbon fibers also showed high oxidation resistance. Thus, lignin-based linear and interconnected carbon fibers exhibited ultrafast charge-discharge and excellent energy density and cyclability performance to be used as binderless and flexible electrodes in supercapacitors.[4] Electrospinning technique allowed preparing lignin-derived carbon fibers with well dispersed metal nanoparticles (Pt) on their surface[5] that showed outstanding catalytical behavior in the electro-oxidation of methanol and ethanol.[6]

Conclusions

A versatile technique to prepare carbon material from and abundant and renewable polymer (lignin) has been reported. The phosphorus functionalization enables to shorten the carbon fibers preparation time, promotes the chemical activation of lignin fibers and greatly increases their oxidation resistance. The resulting carbon materials meet different interesting properties, such as sub-micron metric diameters (≤ 1 μm), large surface area (~2000 m^2/g), relatively high performance in relation to their mechanical properties for functional applications and a rich variety of uniformly-distributed O and P surface functionalities, which make them very attractive for heterogeneous catalysis, adsorption and electrochemical applications.

Acknowledgment

We gratefully thank MINECO and FEDER (Project CTQ2015-68654-R) for financial support.

References

1. Lallave, M., Bedia, J., Ruiz-Rosas, R., Rodríguez-Mirasol, J., Cordero, T., Otero, J.C., Marquez, M. Barrero, A., Loscertales, I.G. (2007). Filled and hollow carbon nanofibers by coaxial electrospinnign of Alcell lignin without binder polymer. *Adv. Mater.* 19, 4292-4296.

2. García-Mateos, F.J., Berenguer, R., Valero-Romero M.J., Rodríguez-Mirasol, J., Cordero, T. (2018). Phosphorus functionalization for the rapid preparation of highly nanoporous submicron-diameter carbon fibers by electrospinning of lignin solutions. *Journal of Material Chemistry A,* 6, 1219-1233

3. Calvo-Muñoz, E.M., García-Mateos, F.J., Rosas, J.M., Rodríguez-Mirasol, J., Cordero, T. Biomass waste carbon materials as adsorbent for CO_2 capture undes post-combustion conditions. (2016). *Frontiers in Materials* 3, 23.

4. Berenguer, R., García-Mateos, F.J., Ruiz-Rosas, R., Cazorla-Amorós, D., Morallón, E., Rodríguez-Mirasol, J., Cordero, T. (2016) Biomass-derived binderless fibrous carbon electrodes for ultrafast energy storage, *Green Chem.* 18, 1506–1515

5. Ruiz-Rosas, R., Bedia, J., Lallave, M., Loscertales, I.G., Barrero, A., Rodríguez-Mirasol, J., Cordero, T. (2010). The production of submicron diameter carbon fibers by electrospinnign of lignin. *Carbon* 48, 696-705.

6. García-Mateos, F.J., Cordero-Lanzac, T., Berenguer, R., Morallón, E., Cazorla-Amorós, D., Rodríguez-Mirasol, J., Cordero, T. (2017) Lignin-derived Pt supported carbon (submicron)fiber electrocatalysts for alcohol electro-oxidation. *Applied Catalysis B: Environmental.* 211, 18-30

Spanish Carbon Group (GEC)
Instituto de Carboquimica, ICB-CSIC
C/ Miguel Luesma Castan
4 · 50018 · Zaragoza · Spain

ISBN 978-1-51087885-3
9 781510 878853

CURRAN ASSOCIATES INC.
proceedings.com

Fachtagung Bauwerksdiagnose (Bau-2024)

Berichtsband BB 181

Berlin, Germany
29 February – 1 March 2024